Revolutionising Crop Management Practices for Ensuring Food Security

NIPA® GENX ELECTRONIC RESOURCES & SOLUTIONS P. LTD.
New Delhi-110 034

About the Editors

Mr. Sumit Sow is currently pursuing Ph.D. in Agronomy with an ICAR-SRF at Dr. Rajendra Prasad Central Agricultural University, Pusa, Bihar. He completed his graduation from Visva-Bharati, West Bengal and his M.Sc. in Agronomy with an ICAR-JRF Scholarship from Bihar Agricultural University, Sabour, Bihar. He has qualified for ICAR-NET and UGC-NET. He has authored two textbooks and four edited books. Additionally, he has published 12 research articles, 7 review articles, 35 popular articles, and 25 book chapters. He has received prestigious awards such as Best Poster, Best Article and Best M.Sc. Thesis Award.

Ms. Shivani Ranjan is a Research Scholar at Dr. Rajendra Prasad Central Agricultural University, Pusa, Bihar. She has completed her B.Sc. (Ag.) from Dr. RPCAU, Pusa, Bihar and M.Sc. in Agronomy from BAU, Sabour, Bihar. She has qualified ICAR-NET, UGC-NET and is a recipient of NFOBC fellowship. She is one of the authors of the book "Extract of Agronomy" and the editor of four books. She has published 11 research articles, 9 review articles, 34 popular articles and 20 book chapters. She has also received various national awards like Best Article and Best M.Sc. Thesis award.

Dr. Dhirendra Kumar Roy is the Director Seeds and a Professor-cum-Chief Scientist at Dr. Rajendra Prasad Central Agricultural University, Pusa. He has also held the esteemed positions of Deputy Director (Pusa farm), Head of various regional research stations, and Officer-in-charge (Kalyanpur farm). He has guided a number of M.Sc. and Ph.D. students. Additionally, he has completed two prestigious externally funded projects and six consultancy projects. He has published 70 research papers, 32 popular articles, 10 technical bulletins, 4 extension bulletins, 6 pamphlets, 3 booklets, 5 folders, 10 book chapters, 4 books, and 1 practical manual.

Dr. Navnit Kumar has been working as an Associate Professor-cum-Sr. Scientist since 2019 in the Department of Agronomy at the Sugarcane Research Institute, Dr. Rajendra Prasad Central Agricultural University, Pusa, Bihar. He completed his B.Sc. (Ag.), M.Sc. (Ag.) and Ph. D degrees from RAU, Pusa, Bihar. With over 18 years of experience in extension, teaching and research, Dr.Kumar has guided 9 students as major advisor, 15 students as co-advisor. He has published 65 research papers in national and international journals, 81 popular articles in farmer friendly magazines, 30 book chapters, 7 authored books, 6 edited books, and 13 folders. He has received various awards for his contributions to teaching and research in the field.

Dr. Rajan Kumar is an Associate Professor-cum-Sr. Scientist in the Department of Agronomy at Dr. Rajendra Prasad Central Agricultural University, Pusa, Bihar. He obtained his Doctoral Degree from Dr. RPCAU, Pusa, Bihar with a Rajiv Gandhi National Fellowship from UGC, New Delhi. With over 14 years of research and teaching, he has supervised eight post graduate students and held various administrative positions. He has published more than 25 research papers, 6 books, 6 book chapters and several popular articles in reputable peer-reviewed journals.

Revolutionising Crop Management Practices for Ensuring Food Security

Editors
Sumit Sow
Shivani Ranjan
Dhirendra Kumar Roy
Navnit Kumar
Rajan Kumar

NIPA® GENX ELECTRONIC RESOURCES & SOLUTIONS P. LTD.
New Delhi-110 034

NIPA® GENX ELECTRONIC
RESOURCES & SOLUTIONS P. LTD.

101,103, Vikas Surya Plaza, CU Block
L.S.C.Market, Pitam Pura, New Delhi-110 034
Ph : +91 11 27341616, 27341717, 27341718
E-mail: newindiapublishingagency@gmail.com
www: www.nipabooks.com

For customer assistance, please contact
Phone: + 91-11-27 34 17 17
Fax: + 91-11-27 34 16 16

Print ISBN: 978-93-58878-55-4

ebook ISBN: 978-93-58877-15-1

Preface

In the ever-evolving landscape of agriculture, the quest for sustainable and efficient crop management practices has become paramount. We are delighted to publish our book entitled “Revolutionising Crop Management Practices for Ensuring Food Security”. In an era marked by environmental challenges, shifting climates, and a growing global population, the need for innovative and resilient crop management practices has never been more pressing. This book brings together a diverse array of insights, methodologies, and case studies that collectively illuminate the path toward ensuring food security for present and future generations.

This book brings together a diverse array of perspectives, methodologies, and case studies that illuminate the path toward sustainable and resilient crop management. It is born out of the conviction that sustainable agriculture is not a choice but a responsibility, a commitment to future generations and the planet we call home. We will delve into topics ranging from soil health and fertility management to precision agriculture, crop rotation, and integrated pest management. The aim is not only to impart knowledge but to inspire a profound sense of stewardship and a vision for an agriculture that respects the Earth's capacity for regeneration.

The chapters in this edited book delve into multifaceted aspects of crop management, addressing issues of precision agriculture, climate-smart practices, technological interventions, and the integration of traditional wisdom with cutting-edge innovations. Each contribution is a unique piece of the puzzle, contributing to the broader picture of how we can ensure not only an abundance of crops but also the conservation of natural resources and the mitigation of environmental impacts. From soil health to pest management, from the use of advanced technologies to indigenous wisdom, the content of this book reflects the holistic approach necessary for a comprehensive transformation in crop management practices.

We are thankful to NIPA, New Delhi for their interest and effort related to the publication of the book.

We hope that the compilation in this book inspires policymakers, researchers, farmers, and all stakeholders involved in agriculture to embrace innovation and collaboration for a future where food security is not just a goal but a sustainable reality.

Editors

Contents

Preface ..... *vii*

1 Role of Artificial Intelligence in Crop Production: Revolutionizing Agriculture for a Sustainable Future 1
Gaurav Verma, Amit Kumar, Adarsh Kumar, Navish Kumar Kamboj, Saurav Saha, T.L. Bhutia, Sonam Ongmu Bhutia, Janak Kumar Singh Sumit Sow and Shivani Ranjan

2 Biofortification: A New Way to Achieve Food and Nutritional Security 13
Sharad Chenalwad, Pundlik Waghmare, Anil Chavan and Santosh Shinde

3 Crop Modeling and Simulation for Climate Smart Crop Production 23
Pundlik Waghmare, Sharad Chenalwad, Santosh Shinde and Anil Chavan

4 Government Initiatives for Sustainable Agriculture in India 35
Anil Chavan, Pundlik Waghmare, Sharad Chenalwad and Santosh Shinde

5 Innovative Management Practices for Diseases and Pests in Field Crops 51
Shashi Kant Ekka, Reena Kujur and Johnson Lakra

6 Interactions Between Agricultural and Environmental Policies 59
Shirisha Junuthula and Veenita Kumari

7 Indigenous Agricultural Practices for Sustainable Crop Production ... 73
Parveen Kumar and Raj Kumar Gupta

8 Legume Cultivation in India Challenges and Future Prospects 83
S. Ashokh Aravind, A. Surendhar, K. Arun Kumar, A. Ammaiyappan and K. Nagarajan

9 Benefits of Millets Cultivation in India 101
Nirmul Kumar, Aditya Bharti and Kumari Sristi

10 Nematode Associated Molecular Pattern (NAMP) Mediated Immune Response Against Plant Parasitic Nematodes......................113

Shandeep S.G., Arun A., Haran M.S.R., Deeikshana T. and Vijay S.

11 Persistence of Herbicides in Crops and Its Management..................125

Dhamodharan P., K. Kiruthika, M. Raghuramakrishnan, G. Gunalarasi and S. Somasundaram

12 Precision and Data-Driven Farming for Sustainable Agriculture......137

Santosh Shinde, Pundlik Waghmare, Sharad Chenalwad and Anil Chavan

13 Recent Advances in Management of Problematic Soil......................149

Reena Kujur, Shashi Kant Ekka, Heba I. Khan, Eugenia P. Lal and Ram Bharose

14 Recent Innovations in Nutrient Management for Maximising Crop Production..159

Souvik Sadhu, Rajni Prabha Rani, Koushik Barik, Dinabandhu Samanta and Ritwik Sahoo

15 Regenerative Practices for Restoring the Soil Fertility.....................177

Sumit Kumar, Rishikesh Yadav, Praveen Kumar Yadav and Sone Lal Patel

16 Water Conservation Techniques for Sustainable Crop Production....195

Bibek Ishore and Tanmoy Das

17 Weather Variability and Crop Yield Forecasting for Sustainable Agriculture..205

Hemareddy Thimmareddy, Nikhil S. Paschapur, A.C. Mahadevaprasad, Vishwanath V. Meti, R. Divya and Yashwanth H.R.

Index...215

1

Role of Artificial Intelligence in Crop Production: Revolutionizing Agriculture for a Sustainable Future

Gaurav Verma[1], Amit Kumar[2], Adarsh Kumar[3], Navish Kumar Kamboj[1], Saurav Saha[2], T.L. Bhutia[2], Sonam Ongmu Bhutia[2] Janak Kumar Singh[2], Sumit Sow[4] and Shivani Ranjan[4]

[1]*Chaudhary Charan Singh Haryana Agricultural, University, Hisar, Haryana 125001 India*

[2]*ICAR Research Complex for NEH Region, Sikkim Centre, Tadong, Gangtok, Sikkim 737102, India*

[3]*ICAR-National Bureau of Agriculturally Important Microorganisms, Mau 275103 Uttar Pradesh, India*

[4]*Department of Agronomy, Dr. Rajendra Prasad Central Agricultural University Pusa 848125, Bihar, India*

Abstract

Agriculture, constituting 6.4 percent of the global GDP, serves as the primary driver for both food production and economic output. It is a significant source of revenue in nine countries across the globe. The field of digital agriculture management applications has seen remarkable advancements, leveraging information and communication technology (ICT) to deliver benefits for farmers and consumers alike, extending technological solutions into rural landscapes. This chapter underscores the potential of ICT technologies in traditional agriculture, exploring the advantages they bring while also addressing the challenges inherent in their integration into farming practices. The document delves into issues surrounding robotics, Internet of Things (IoT) devices, and machine learning, elucidating the roles of artificial intelligence and sensors in the agricultural domain. Drones are being increasingly considered for crop observation and optimising crop yield management. Furthermore, the review highlights global and cutting-edge IoT-based farming systems and platforms where applicable. It concentrates on recent technological strides in remote sensing and AI, emphasising their role in enhancing the resilience of agricultural systems. In essence, this exploration signifies the transformative potential of digitalisation and advanced technologies in shaping the future of sustainable and resilient agriculture.

Keywords: *Artificial intelligence, Digitalisation, IoT, ICT, Resilient agriculture*

Introduction

Agriculture has been the cornerstone of human civilisation for millennia, providing sustenance and livelihoods to communities around the world. As the global population continues to grow, reaching an estimated 9.7 billion by 2050, the need to ensure food security becomes increasingly urgent. However, this burgeoning demand for food coincides with numerous challenges such as climate change, dwindling natural resources, and the pressures of sustainable farming practices. To address these complex issues and meet the world's food requirements sustainably, modern agriculture must embrace innovative technologies. One such game-changing technology that has emerged in recent years is Artificial Intelligence (AI), which is revolutionising various industries, including crop production. AI is the simulation of human intelligence processes by machines, particularly computer systems. It encompasses a range of capabilities, including learning, reasoning, problem-solving, perception, and language understanding. When applied to agriculture, AI holds immense promise to transform how we grow crops, optimise resource utilisation, and enhance overall productivity (Fan *et al.*, 2012). By leveraging AI's power, farmers can make data-driven decisions, improve efficiency, and implement precision farming techniques tailored to specific field conditions. By employing AI technologies, it is possible to mitigate the impact on natural ecosystems and enhance worker safety. This, in turn, can contribute to keeping food prices down and ensuring that food production remains in step with the growing global population (Ranjan *et al.*, 2023).

Challenges in Agriculture using Traditional Methods

Before we can grasp the influence and use of AI in agriculture, we must first understand the obstacles that farmers face while utilising traditional methods, which are listed below:

- Different meteorological conditions; such as rainfall, temperature, and humidity have a vital influence on farming. Due to climate change, traditional methods and old observatories are unable to provide accurate forecasting resulting in making false judgements and creating problems in field preparation, sowing of seeds, intercultural operations, harvesting, storage of produce etc.

- To achieve good crop yield, crops require nutrients in abundant amounts. The crop's need can be met through different organic and inorganic sources. However, under field conditions, there is a high fertility gradient resulting in uneven availability of nutrients to the crop. Even uniform application of nutrients is not helpful in enhancing fertiliser use efficiencyi.e. a limitation of the traditional method. The problem can be sorted out with the use of AI, which and it will help us attain higher yields along with an increase in input use efficiencies.
- Weeds areanother limitation in crop production. They are hardy in nature, have a high multiplication rate, and can grow widely. Weeds compete with the crop for nutrients, space, water, sunlight etc., limiiting crop growth. Traditional methods of weed management are labour-intensive and not cost-effective. Even the application of herbicides has resulted in the development of herbicide resistance in weeds. The application of AI in weed management can help us reduce the incidence of herbicide resistance development and be cost-effective.

What is Artificial Intelligence?

Artificial intelligence (AI) is currently at the forefront of human technology. AI and automation have already influenced industries such as healthcare, banking, automotive, retail, and manufacturing (Mueller and Massaron, 2018).

AI systems must exhibit characteristics associated with human intellect. Reasoning, planning, learning, problem-solving, knowledge representation, and other activities are examples of these behaviours. Because of machine learning, AI may grow indefinitely (Vuppalapati, 2021).

Approach Utilizing AI and Machine Learning

The application of AI involves leveraging technology to replicate human intelligence and processes, encompassing computer systems, robotics, and digital equipment. AI finds diverse applications, including natural language processing (NLP) for understanding spoken communication, computer vision for analog-to-digital conversion like video, and expert systems for simulating judgment. The cognitive skills underlying AI include learning (acquiring and transforming data into actionable information), reasoning (selecting the right algorithm for desired outcomes), and self-correction (continuously adjusting algorithms for accuracy) (Gharaei *et al.*, 2019).

Machine learning (ML), a central aspect of AI, employs statistical and mathematical methods to make data-driven predictions and decisions. ML encompasses symbolic approaches (explicit representation of induced rules and examples) and sub symbolic approaches (such as artificial neural networks: ANN). ML tasks include supervised, unsupervised, and reinforcement learning. Supervised learning involves mapping variables to a preferred output variable using labeled data, employing algorithms like decision trees, Bayesian networks, and regression analysis. Unsupervised learning includes techniques like clustering, genetic algorithms, and deep learning, utilizing unlabeled datasets for pattern recognition and exploratory data analysis (Jordan and Mitchell, 2015). Reinforcement learning employs algorithms like Q-learning and deep Q-learning for skill acquisition, robot navigation, and real-time decision-making, addressing an exploration versus exploitation dilemma (Traore *et al.*, 2017).AI technologies have recently found application in the agri-food sector, offering contributions to model identification, service creation, and decision-making processes. The primary goal is to provide precision and forecasting decisions, enhancing productivity with resource preservation. AI tools, utilising algorithms, evaluate performance, classify patterns, and predict problems in agriculture. Remote sensing and sensors assess abiotic and biotic factors for optimising crop and livestock management (Suprem *et al.*, 2013).

The advantages of AI implementation in the agri-food sector are manifold. It improves efficiency in crop production, harvest, and sales, emphasises defect checking, enhances potential for healthy crop production, and aids in weather forecasting and disease or pest identification with high accuracy (Sujatha *et al.*, 2021). AI techniques support agro-based businesses by addressing challenges like climate variation and pest infestation, as demonstrated by Crane-Droesch's novel modeling approach (Crane-Droesch, 2018). AI tools provide real-time weather data, allowing farmers to make informed decisions, as evidenced by Fente and Singh's weather forecasting model. AI approaches monitor soil health, detecting defects and nutrient deficiencies through image recognition or deep learning (Suchithra and Pai, 2020). Significantly, AI technology contributes to environmental protection by reducing pesticide usage. Robotics, computer vision, and machine learning help farmers target chemical spraying only where needed, minimising environmental impact. Additionally, advanced AI-based technologies offer benefits in the agri-food supply chain, including cost reduction, time efficiency, error reduction, decreased human intervention, and automated decision-making (Kamilaris and Prenafeta-Boldú, 2018).

Application of AI in Agriculture

AI is transforming agriculture by providing creative solutions to increase production, sustainability, and efficiency. Here are some of the most important uses of AI in agriculture (Ahamed, 2023).

1. Crop monitoring: Early detection of crop diseases and pest infestations is critical to preventing their spread and mitigating potential losses. AI-enabled image recognition systems can monitor crops continuously, identifying subtle changes in plant health and alerting farmers to potential issues. By leveraging machine learning, these systems become more accurate over time, refining their identification capabilities and allowing farmers to respond proactively. Consequently, AI aids in reducing the dependence on chemical treatments, promoting sustainable agriculture. AI-powered drones equipped with hyperspectral cameras can capture images of crops from above, detecting variations in chlorophyll content and other indicators of plant health. By analysing these images, AI algorithms can identify potential disease outbreaks before they become visually apparent, enabling timely intervention and minimising crop losses.
2. Precision farming: Precision farming is a key area where AI is driving significant advancements in crop production. AI-powered systems gather data from various sources, such as drones, satellites, weather stations, and sensors on the ground. This data is then processed using AI algorithms to create precise maps of fields, enabling farmers to understand variability in soil conditions, moisture levels, and crop health. By providing real-time information, AI helps farmers optimise irrigation schedules, tailor fertiliser applications, and manage pests more effectively. This targeted approach minimises resource wastage and enhances overall crop yields. The integration of AI and precision agriculture empowers farmers to make data-driven decisions. Instead of relying on traditional blanket treatments across entire fields, farmers can now adapt their strategies to the specific needs of individual areas. This reduces unnecessary chemical use and mitigates environmental impacts, fostering sustainable agricultural practices.
3. Autonomous farming:AI-driven robotics and autonomous machinery are making great strides in crop production. These intelligent machines can perform tasks such as planting, harvesting, and weeding with unparalleled accuracy and speed. They can work tirelessly in adverse weather conditions and offer greater flexibility in scheduling farm operations. As a result, farmers can improve operational efficiency, reduce labor costs, and optimise the use of inputs, leading to increased productivity and reduced environmental impact. For example, AI-powered harvesting machines equipped with computer vision and machine learning algorithms can identify ripe fruits and vegetables,

ensuring precise and efficient harvesting. This minimizes food waste and labor requirements while maximising the quality and yield of the harvest.

4. Livestock monitoring: Using sensors and computer vision, AI-based systems can monitor the health and behaviour of animals. They can identify illness early on, monitor eating habits, and optimise breeding operations. This assists farmers in improving animal welfare, illness prevention, and production.
5. Optimization of supply chain:By analysing aspects such as demand, transit routes, and storage conditions, AI algorithms can optimise supply chain logistics. This improves inventory management, lowers waste, and ensures that agricultural goods are delivered on time.
6. Yield prediction:To anticipate agricultural yields, AI models can analyse historical data, weather patterns, and other pertinent aspects. This data helps farmers make more educated decisions about planting, harvesting, and market planning, resulting in increased efficiency and profitability.
7. Forecasting of disease and pests:AI algorithms can forecast disease outbreaks and pest infestations by analysing historical data, weather patterns, and other relevant variables. Early identification and intervention assist farmers in taking preventative steps, minimising crop loss, and reducing the need for chemical treatments.
8. Management of soil health:By analysing soil samples and sensor data, AI-based systems can monitor soil health. Farmers may use this to better assess soil conditions, nitrogen levels, and potential soil degradation concerns. Farmers can use this knowledge to apply targeted soil management practices and improve soil health.
9. Agri-Fintech: Farmers can make educated decisions thanks to AI-powered platforms and applications that provide financial and market analytics. These systems analyse market trends, crop performance, and financial data to offer personalised recommendations as well as access to loans and insurance.
10. Artificial intelligence in self-driving tractors: The self-driving tractors can execute a variety of tasks without the need for human intervention. While travelling at the optimal pace, these tractors will avoid impediments and independently recognise their ploughing location. Without worrying about driver weariness, these tractors can do more jobs and manage various operations without tiring.

Table 1. Use of different AI models in crop management (Eli-Chukwu, 2019)

Applications	**Method/Models**	**Crop**	**Input**
High Throughput Phenotyping (HTP)	Deep convolution neural network (DCNN)	Wheat, Soyabean	Genotypic and phenotypic data data
Yield prediction	Deep neural network (DNN), Artificial neural network (ANN), Random forest (RF)	Wheat, Soyabean	Unmanned Aerial images (RGB, Multispectral, Thermal)
Detection crop nutrition disorder	ANN	Rice, Cotton	RGB images, Genotypic and phenotypic data data
Weed detection	Fully convolutional network (FCN)	Rice	UAS images (RGB)
Disease detection	CNN, DCNN, RF	Bean, Maize and Wheat	Field images
Biomass	RF	Barley	UAS images
Soil texture and temperature	ANN	Rice, Maize	UAS images

Advantages of AI in Agriculture

- Decision making: Predictive analytics is a huge boon to the agriculture business. It assists farmers in addressing critical farming difficulties such as market demand analysis, price forecasts, and determining the best dates for crop sowing and harvesting. Furthermore, AI-powered robots can assess soil and crop health, provide fertiliser suggestions, monitor the weather, and assess crop quality. All of these advantages of AI in agriculture help farmers make better decisions and farm more efficiently. AI helps bridge the gap between genomics and phenomics in crop production (Fig. 1).

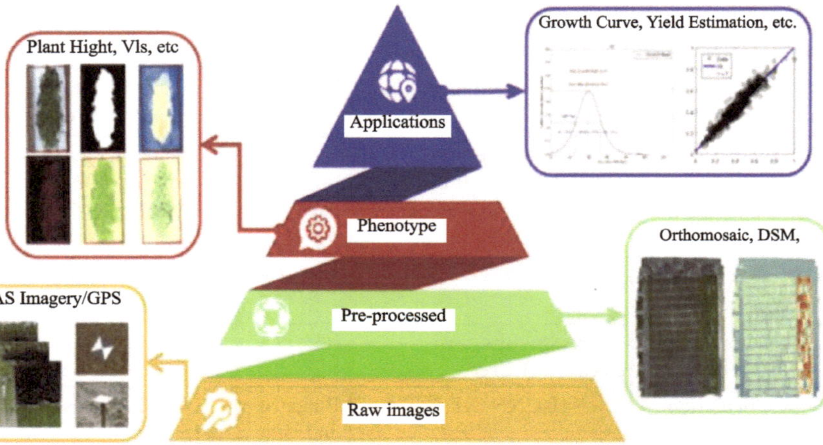

Fig. 1: AI based HTP application development workflow. (VIs-Vegetation Indices, GPS-Global Positioning System, DSM-Digital Surface Model)
Source: Jung *et al.*, 2021

- Cost saving:Precision farming with AI-enabled technology allows farmers to grow more crops with fewer resources and at a lower cost. AI provides farmers with real-time insights, enabling them to make informed decisions at each step of farming. This accurate judgement results in reduced product and chemical waste and more efficient use of time and money. Furthermore, it enables farmers to identify specific regions that require irrigation, fertilisation, and pesticide treatment, thereby reducing the usage of chemicals on the crop. All of this adds up to less herbicide usage, higher crop quality, and more profit with fewer resources.
- Labour saving:In the agricultural enterprise, there has always been a workforce deficit. With farming automation, AI can overcome this problem. Farmers can get more work done using AI and automation, such as driverless tractors, smart irrigation and fertilising systems, smart spraying, vertical farming software, and AI-based harvesting robots. When compared to human farmhands, AI-driven robots and equipment are significantly faster and more precise.

Challenges of AI

Given the benefits of AI for sustainable farming, using this technology may appear to be a sensible move for any farmer. However, several severe problems in AI adaptation are as follows (Satapathy *et al.*, 2022):

- Unawareness among the farmers about artificial machines: Despite the numerous advantages of employing AI in agriculture, most people across the world are unfamiliar with the usage of AI-enabled solutions and equipment. To address the challenges, AI businesses should provide farmers with basic equipment and, once they are comfortable with it, introduce sophisticated machinery.
- Lacking of experience with emerging technologies: Adoption of AI and new technologies in agriculture might be difficult for underdeveloped countries. It will be extremely challenging to sell such technologies in places where no such agricultural equipment is being used. Farmers in such areas require assistance in order to adopt this technology.
- Privacy and security issue: Because there are no defined rules and standards in place for deploying AI yet, it may generate a number of legal difficulties. Furthermore, when software and the internet are used, there may be privacy and security risks, such as hacks and data breaches. All of these difficulties might pose a significant challenge to farm owners or farmers.

Future Prospects and Policy Implication of AI

While the practical implications are primarily associated with technological advancements and the implementation of new business models, notable policy implications also arise. These policies can be tied to the identified barriers in the practical application of new AI-based solutions, encompassing challenges such as the lack of farmers' ICT knowledge and dynamics related to technology acceptance. It is observed that, in some cases, these barriers are specific to agriculture, particularly concerning the complexity and lack of integration in the food supply chain. However, the majority of barriers are general and common across various sectors.As previously suggested, the agricultural sector has the potential to adopt or adapt solutions already created and implemented in other sectors, effectively addressing a significant number of challenges. Policymakers are encouraged to facilitate collaboration among key agricultural stakeholders and actors from diverse sectors to address these general implementation barriers. Governments play a crucial role in promoting the development of new solutions and the adaptation of existing systems. This may involve offering dedicated funds or tax privileges to support farmers, especially small-sized companies, in acquiring technology.

Initiatives focused on knowledge translation and dissemination, engaging multiple stakeholders such as agricultural consortia, technology providers, research institutes, and universities, can contribute to overcoming acceptance issues These efforts aim to enhance understanding of new opportunities for individual farms and the broader ecosystem. In essence, fostering collaboration and creating a supportive policy environment is vital for overcoming barriers and realising the potential benefits of AI-based solutions in agriculture.

Conclusion

Artificial Intelligence is revolutionising crop production, driving a paradigm shift towards precision, sustainability, and productivity. With AI-powered precision farming, automated machinery, crop monitoring, and climate prediction, agriculture can become more efficient and environmentally friendly. By harnessing the potential of AI, farmers worldwide can overcome the challenges posed by a growing population and a changing climate. However, to fully realise the benefits, collaborative efforts from policymakers, researchers, and agricultural stakeholders are crucial in promoting AI adoption and ensuring a more prosperous and sustainable future for global agriculture.

References

Abraham, A., Dash, S., Rodrigues, J. J., Acharya, B., and Pani, S. K. (Eds.). (2021). AI, Edge and IoT-Based Smart Agriculture. Academic Press.

Ahamed, T. (Ed.). (2023). IoT and AI in Agriculture: Self-sufficiency in Food Production to Achieve Society 5.0 and SDG's Globally. Springer Nature.

Crane-Droesch, A. (2018). Machine learning methods for crop yield prediction and climate change impact assessment in agriculture. Environmental Research Letters, 13(11), 114003.

Eli-Chukwu, N. C. (2019). Applications of Artificial Intelligence in Agriculture: A Review. Engineering, Technology & Applied Science Research, 9(4), 4377-4383.

Fan, M., Shen, J., Yuan, L., Jiang, R., Chen, X., Davies, W. J., and Zhang, F. (2012). Improving crop productivity and resource use efficiency to ensure food security and environmental quality in China. Journal of Experimental Botany, 63(1), 13-24.

Gharaei, A., Karimi, M., and HoseiniShekarabi, S. A. (2019). An integrated multi-product, multi-buyer supply chain under penalty, green, and quality control polices and a vendor managed inventory with consignment stock agreement: the outer approximation with equality relaxation and augmented penalty algorithm. Applied Mathematical Modelling, 69, 223–254.

Jordan M. I., and Mitchell, T. M. (2015). Machine learning: trends, perspectives, and prospects. Science, 349.

Jung, J., Maeda, M., Chang, A., Bhandari, M., Ashapure, A., and Landivar-Bowles, J. (2021). The potential of remote sensing and artificial intelligence as tools to improve the resilience of agriculture production systems. Current Opinion in Biotechnology, 70, 15-22. doi:10.1016/j.copbio.2020.09.003

Kamilaris, A., and Prenafeta-Boldú, F. X. (2018). Deep learning in agriculture: a survey. Computers and Electronics in Agriculture, 147, 70–90.

Mueller, J. P., and Massaron, L. (2018). Artificial Intelligence for Dummies. Newark: For Dummies.

Ranjan, S., Sow, S., Kumar, M., Kumar, N., Padhan, S. R., Roy, D. K., Dayal, P., Nath, D., Kumar, P., and Gitari, H. (2023). Artificial Intelligence for Maximizing Agricultural Input Use Efficiency: Exploring Nutrient, Water and Weed Management Strategies. Front. Environ. Sci., 11. doi: 10.3389/fenvs.2023.1243292

Satapathy, S., Mishra, D., Realyvásquez Vargas, A., Satapathy, S., Mishra, D., and Realyvásquez Vargas, A. (2022). Food Safety and IoT-Based Solution. Innovation in Agriculture with IoT and AI, 79–98.

Suchithra M. S., and Pai, M. L. (2020). Improving the prediction accuracy of soil nutrient classification by optimising extreme learning machine parameters. Information Processing in Agriculture, 7(1), 72–82.

Sujatha, R., Chatterjee, J. M., Jhanjhi, N., and Brohi, S. N. (2021). Performance of deep learning vs machine learning in plant leaf disease detection. Microprocessors and Microsystems, 80, 103615.

Suprem, A., Mahalik, N., and Kim, K. (2013). A review on application of technology systems, standards and interfaces for agriculture and food sector. Computer Standards & Interfaces, 35(4), 355–364.

Traore, B. B., Kamsu-Foguem, B., and Tangara, F. (2017). Data mining techniques on satellite images for discovery of risk areas. Expert Systems with Applications, 72, 443–456.

Vuppalapati, C. (2021). Machine Learning and Artificial Intelligence for Agricultural Economics: Prognostic Data Analytics to Serve Small Scale Farmers Worldwide (Vol. 314). Springer Nature.

2

Biofortification: A New Way to Achieve Food and Nutritional Security

Sharad Chenalwad, Pundlik Waghmare
Anil Chavan and Santosh Shinde

Department of Agronomy, Vasantrao Naik Marathwada Krishi Vidyapeeth Parbhani (MS), India

Abstract

Biofortification is a transformative approach in agriculture aimed at enhancing the nutritional value of food crops by increasing the concentration of essential vitamins and minerals in their edible parts. This innovative method addresses food and nutritional security by fortifying staple crops with vital nutrients, contributing to improved health outcomes, particularly in regions where access to diverse diets and nutritional supplements is limited. Through selective breeding, conventional plant breeding techniques, or biotechnology, biofortification elevates the levels of key micronutrients like iron, zinc, vitamin A, and other essential vitamins and minerals in crops such as rice, wheat, maize, beans, and sweet potatoes. By enhancing the nutritional content of these commonly consumed foods, biofortification offers a sustainable solution to combat hidden hunger or micronutrient deficiencies prevalent in many populations globally. This agricultural strategy presents a promising pathway to alleviate malnutrition, especially in vulnerable communities where diets heavily rely on staple crops lacking in essential nutrients. By integrating biofortified crops into existing agricultural systems, it provides an accessible and cost-effective means of delivering crucial vitamins and minerals to populations who might otherwise lack access to diverse food sources or nutritional supplements. Biofortification aligns with broader efforts to address global food and nutrition challenges sustainably. By leveraging advancements in agricultural science and technology, it offers a powerful tool to enhance food and nutritional security, ultimately improving public health and well-being worldwide.

Keywords*: Agronomical fortification, Food security, Nutritional security, Sustainable agriculture*

Introduction

Biofortification, also known as biological fortification, involves enhancing the nutritional content of food crops to improve bioavailability for the human population. This is achieved through the use of modern biotechnological techniques, conventional plant breeding, and agronomic practices. Unlike our current agricultural system, which primarily focuses on increasing grain yield and crop productivity, biofortification addresses the nutritional needs of the human population. The prevailing approach has led to a rapid increase in micronutrient deficiency in food grains, contributing to higher rates of micronutrient malnutrition among consumers. Biofortification presents a one-time investment that offers a cost-effective, long-term, and sustainable solution to combat hidden hunger (Sow and Ranjan, 2021a). Once biofortified crops are developed, there are no ongoing costs associated with purchasing fortificants and adding them to the food supply during processing. Agronomic biofortification methods involve the physical application of nutrients to temporarily enhance the nutritional and health status of crops. The consumption of such crops subsequently improves human nutritional status. Successful implementation of the biofortification initiative requires farmers to receive support for using micronutrient fertilisers on fortified cultivars and marketing assistance. Despite challenges, the biofortification of millets holds a promising future in addressing the issue of malnutrition.

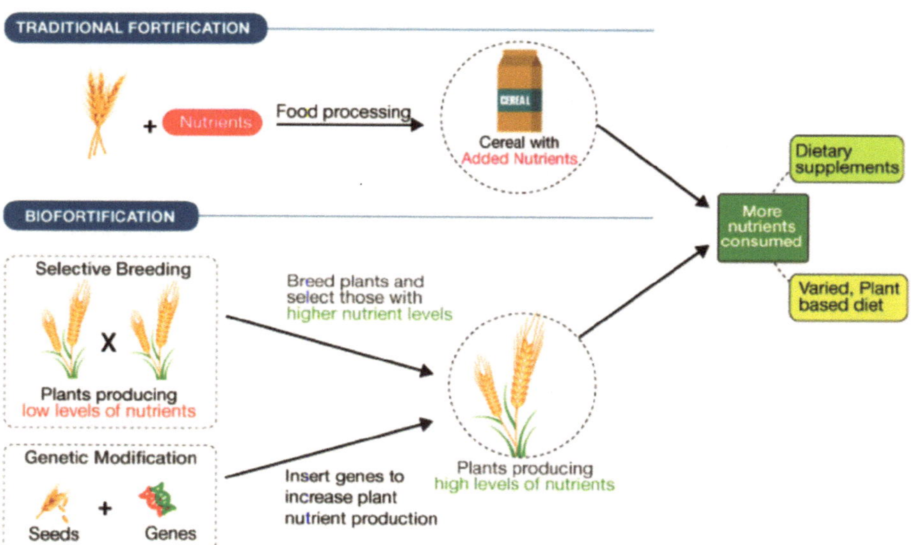

Fig. 1: Biofortification - Improving Nutrition for Reducing Malnutrition

In developing nations, 780 million out of approximately 792.5 million malnourished individuals are located, with around 2 billion people globally experiencing "hidden hunger"—a deficiency in essential micronutrients in their diets. Traditionally, our agricultural focus has been on maximising crop productivity and yield rather than prioritising human health. This approach has led to a decline in the micronutrient content of dietary grains, exacerbating micronutrient malnutrition among consumers. The agricultural landscape is shifting from solely increasing food crop production to emphasising the cultivation of nutrient-rich crops. This transition aims to combat "hidden hunger" prevalent in underdeveloped regions, where diets predominantly consist of micronutrient-deficient staple foods. Despite significant advancements since the Green Revolution, the nutrient content of crops hasn't kept pace with population growth, exacerbating malnutrition due to imbalanced diets, especially in developing countries.

The issue of malnutrition often referred to as "hidden hunger," has gained critical importance in global agendas following the Millennium Development Goals (MDGs) and the subsequent adoption of the Sustainable Development Goals (SDGs). Reports indicate that approximately 2 billion people worldwide suffer from malnutrition, with over 850 million individuals affected by undernourishment. Regions across the world grapple with malnutrition, particularly in low-income countries like Africa, where a significant portion of the population faces risks of micronutrient deficiencies, notably in calcium (54%), zinc (40%), selenium (28%), iodine (19%), and iron (5%) (Das *et al.*, 2013). This deficiency disproportionately affects mothers and young children in these regions. Addressing the prevalence of malnutrition globally requires innovative strategies. Biofortification, involving the enhancement of various crop varieties to contain higher levels of essential nutrients, offers a sustainable, long-term solution. This approach utilises contemporary biotechnology methods, traditional plant breeding, and agronomic practices to produce nutrient-enriched food crops. Biofortified crops, designed with increased bioavailable concentrations of vital micronutrients, reach consumers through existing agricultural and food trade channels. This approach effectively reaches undernourished families and low-income groups with limited access to diverse diets, supplements, or fortified foods. Importantly, these crops offer improved bioavailability, ensuring greater nutrient absorption by the human body.

Biofortification emerges as a promising, cost-effective, and sustainable means of addressing micronutrient deficiencies in populations limited by access to diverse diets and alternative interventions. Agronomic biofortification, a practice enhancing the micronutrient content of food crops through agronomic techniques, offers a swift, safe, and economical approach to augmenting iron, zinc, and other essential micronutrients in our diets. Unlike molecular/genetic methods, agronomic biofortification works on existing crop varieties, ensuring consumer acceptability (Hefferon, 2017).

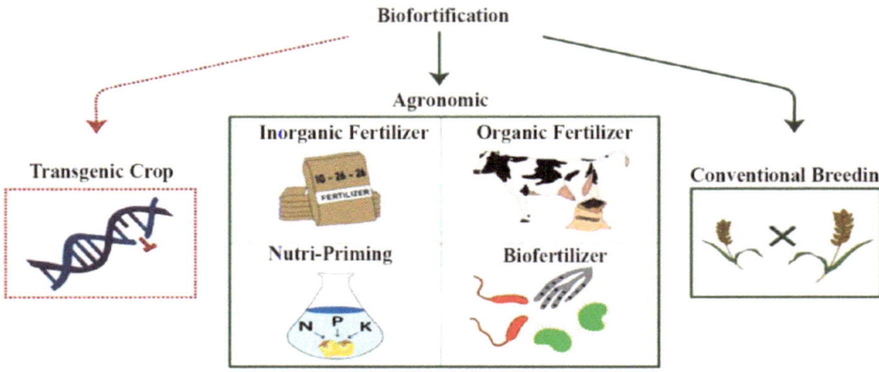

Fig. 2: Methods of biofortification

Notably, major food crops often lack the necessary micronutrients vital for human health. Biofortified food crops, including cereals, legumes, vegetables, and fruits, play a crucial role in delivering these micronutrients to targeted populations. While transgenic research receives significant attention, breeding methods demonstrate higher success rates and acceptance levels. Despite challenges, biofortified crops present a promising future in combating hunger. However, barriers persist, including the purchasing power of underprivileged communities, limited access to markets and healthcare systems, and a lack of awareness about the long-term health benefits of these nutrient supplements.

Biofortified crop development eliminates the need to purchase fortification agents separately and add them during food processing. This economic perspective positions biofortification as a one-time investment, offering a cost-effective, sustainable, and long-term solution to addressing hidden hunger. Looking ahead, the developing world anticipates a substantial population surge in the coming decades, coupled with changing climate conditions. These factors magnify the challenges in ensuring food security. Therefore, biofortification stands as a critical strategy in securing nutritional needs amid these evolving global dynamics (FAO, 2015). Given that a significant portion of the global population adheres to a plant-based diet, ensuring nutritional security is crucial for enhancing overall health. Plants serve as the primary source of essential nutrients required for healthy growth and development. However, a substantial portion of the world's population, particularly in Asia and Africa, experiences nutritional deficiencies primarily due to heavy dependence on grain products.

Biofortified crops are designed to address specific nutrient deficiencies prevalent in certain populations. Several key criteria define the success and effectiveness of biofortified crops:

1. Targeted Nutrient Enhancement: Biofortified crops are developed to enhance specific essential nutrients like iron, zinc, vitamin A, iodine, and others that are deficient in the diets of particular regions or communities (Khush *et al.*, 2012).
2. Retained Agronomic Traits: They maintain the essential agronomic traits of the original crop varieties, ensuring that the biofortified versions are as productive, resilient, and adaptable as their non-biofortified counterparts.
3. Bioavailability: The increased nutrients in biofortified crops should be easily absorbed and utilised by the human body. This emphasises the importance of not just increasing the quantity of nutrients but also ensuring their bioavailability.
4. Consumer Acceptance: The taste, appearance, and cooking qualities of biofortified crops are critical. Consumer acceptance is crucial for the adoption and success of these crops, ensuring that people are willing to incorporate them into their diets.
5. Adaptability to Local Conditions: Biofortified crops should be adaptable to diverse agro-ecological conditions prevalent in different regions. This ensures their viability and effectiveness across various climates and agricultural practices.
6. Sustainability: The production of biofortified crops should be sustainable, with minimal environmental impact and the ability to integrate into existing farming systems without significant alterations.
7. Accessible to Targeted Populations: These crops should be affordable and accessible to the populations most in need of these nutrients, especially those with limited resources or access to diverse diets.
8. Evidence-Based Impact: Scientific research and evidence supporting the efficacy and impact of biofortified crops on reducing micronutrient deficiencies play a vital role in their development and adoption.
9. Safety: Ensuring that biofortified crops meet safety standards and regulations is crucial. They should not pose any health risks or introduce harmful substances into the food chain.

Biofortified crops that meet these criteria have the potential to significantly alleviate micronutrient deficiencies and improve the overall nutritional status of vulnerable populations worldwide (Sow and Ranjan, 2021b).

Furthermore, biofortification has witnessed several notable developments aimed at addressing widespread nutritional deficiencies. Some key advancements include:

1. Breeding Programmes: Significant progress has been made in conventional breeding methods to enhance the micronutrient content of crops. These programs focus on selecting and crossbreeding varieties with naturally higher nutrient levels.
2. Genetic Engineering: Biotechnological advancements have facilitated the development of genetically modified (GM) crops that possess enhanced nutrient content. This involves the insertion of specific genes responsible for producing desired nutrients into the crop's genome.
3. Trait Stacking: Scientists are exploring methods to combine multiple desirable traits in a single crop. This approach involves stacking various genes responsible for different nutrients or agronomic traits, creating multi-fortified crops.
4. Biofortified Crop Varieties: There has been a considerable expansion in the range of biofortified crop varieties. Notably, biofortified rice, maize, wheat, beans, sweet potatoes, and cassava varieties have been developed to address specific micronutrient deficiencies.
5. Bioavailability Studies: Research focuses on improving the bioavailability of nutrients in biofortified crops. This includes identifying factors that influence nutrient absorption and utilization within the human body to ensure maximum nutritional impact.
6. Agronomic Practices: Innovations in agricultural techniques, such as soil and crop management practices, aim to optimise nutrient uptake and enhance the nutrient content of crops without genetic modification.
7. Public-Private Partnerships: Collaborations between governments, research institutions, NGOs, and the private sector have accelerated biofortification initiatives. These partnerships facilitate research, funding, distribution, and scaling up of biofortified crops.
8. Adoption and Scaling: Efforts have been made to promote the adoption and widespread cultivation of biofortified crops. This includes educating farmers and communities about the benefits of these crops and establishing seed distribution programmes.
9. Nutritional Studies: Ongoing research evaluates the impact of biofortified crops on alleviating micronutrient deficiencies. Studies assess their effectiveness in improving nutritional outcomes and reducing the prevalence of related health issues.

These developments underscore the multifaceted approach taken to address nutrient deficiencies through biofortification. Continuous research, technological advancements, and collaborative efforts are key to furthering the impact of biofortified crops on global nutrition and food security.

Significance of Biofortification in Nutritional and Food Security

Biofortification is profoundly significant in addressing nutritional and food security challenges due to its multifaceted impact:

1. Combatting Hidden Hunger: Biofortified crops, enriched with essential nutrients like iron, zinc, vitamin A, and others, directly combat hidden hunger. This form of malnutrition occurs when people consume enough calories but lack essential vitamins and minerals, leading to health issues despite seemingly adequate diets.
2. Improved Nutritional Intake: By increasing the nutritional content of staple crops, biofortification ensures that individuals consuming these staples receive vital nutrients. This is particularly crucial in regions where these crops are dietary mainstays.

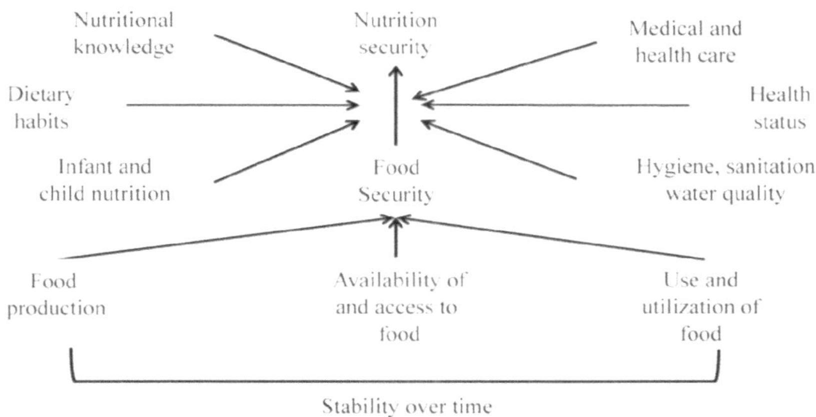

Fig. 3: Significance of biofortification in nutritional and food security

3. Accessibility and Affordability: Biofortified crops offer a sustainable and cost-effective solution to improve nutrition. They integrate essential nutrients into commonly consumed foods, making them accessible to populations who might not have access to diverse diets or supplements.
4. Health Benefits: Enhanced nutrient intake from biofortified foods contributes to improved health outcomes, especially among vulnerable groups like children, pregnant women, and the elderly. It can reduce

the incidence of diseases related to micronutrient deficiencies, such as anemia or vision impairment.

5. Sustainability in Agriculture: Biofortification integrates nutritional enhancement into existing agricultural systems. This sustainable approach aligns with traditional farming practices and reduces reliance on external interventions or supplements.
6. Global Impact: As staple crops are consumed by millions worldwide, the widespread adoption of biofortified varieties has the potential for a substantial global impact, benefiting large populations affected by malnutrition.
7. Resilience to Environmental Challenges: Enhanced nutrient content in biofortified crops can contribute to greater resilience against environmental stresses. Crops fortified with certain nutrients may exhibit improved resistance to pests, diseases, or adverse growing conditions.
8. Economic Benefits: Improved nutrition from biofortified crops can potentially increase productivity and reduce healthcare costs associated with addressing nutrient-deficiency-related illnesses. This can have positive economic implications for communities and nations (Gilani and Nasim, 2007).

In short, biofortification plays a crucial role in addressing both nutritional and food security concerns. By enhancing the nutritional quality of staple crops, it offers a sustainable and cost-effective strategy to combat hidden hunger, improve health outcomes, and contribute to overall well-being, particularly in regions where malnutrition is prevalent.

Conclusion

In conclusion, biofortification stands as a beacon of hope in the fight against malnutrition and food insecurity. Its ability to fortify staple crops with essential nutrients directly targets hidden hunger, ensuring that even the most vulnerable populations receive vital vitamins and minerals through their everyday diets. This innovation not only elevates the nutritional intake of communities heavily reliant on these crops but also enhances health outcomes, especially for children, pregnant women, and the elderly. Moreover, biofortification's sustainable and cost-effective nature aligns with traditional farming practices, offering a long-term solution that reduces dependency on external interventions. Its potential for global impact, bolstering the resilience of crops against environmental stresses while also promising economic benefits by improving productivity and reducing healthcare costs, cannot be overstated. In essence, biofortification is not merely about enriching crops; it's about fortifying communities, improving health, and paving the way for a more nourished and secure future worldwide. This transformative approach in sustainable agriculture holds immense promise, addressing fundamental nutritional needs and shaping a healthier, more resilient world for generations to come.

References

Das, J.K., Kumar, R., Salam, R.A., and Bhutta, Z.A. (2013). Systematic review of zinc fortification trials. Annals of Nutrition and Metabolism, 62(1), 44-56.

FAO. (2015). The state of food insecurity in the world 2015: Meeting the 2015 international hunger targets: Taking stock of uneven progress. Rome: FAO. Advances in Nutrition, 6(5), 623-624.

Gilani, G.S., and Nasim, A. (2007). Impact of foods nutritionally enhanced through biotechnology in alleviating malnutrition in developing countries. Journal of AOAC International, 90(5), 1440-1444.

Hefferon, K.L. (2016). Can biofortified crops help attain food security? Current Molecular Biology Reports, 2(4), 180-185.

Khush, G.S., Lee, S., Cho, J.I., and Jeon, J.S. (2012). Biofortification of crops for reducing malnutrition. Plant Biotechnology Reports, 6, 195-202.

Sow, S. and Ranjan, S. (2021a). A Review on Agronomic Biofortification for Improving Food and Nutritional Security. Research Biotica, 3(2), 139-144.

Sow, S. and Ranjan, S. (2021b). Zinc Biofortification in Groundnut. Agriblossom, 1(11), 55-61.

3

Crop Modeling and Simulation for Climate Smart Crop Production

Pundlik Waghmare, Sharad Chenalwad Santosh Shinde and Anil Chavan

Department of Agronomy, Vasantrao Naik Marathwada Krishi Vidyapeeth Parbhani (MS), India

Abstract

The rising global population necessitates a substantial increase in food production to sustain the growing demand, amid the challenges posed by climate shifts like rising temperatures and altered rainfall patterns. Predicting crop outcomes under diverse climatic scenarios and management strategies becomes pivotal for empowering farmers with adaptable approaches to optimise sustainable crop growth. Crop modeling tools serve as a structured framework to assess potential climate adaptations, forming the basis for farmer-centric decision support systems. Leveraging weather and environmental data, these models simulate crop development, growth, yield, and resource uptake. While the primary goal of crop models is precise economic yield estimation, their complexity varies based on available information, spanning from empirical models using monthly weather variables for regional yield predictions to detailed mechanistic models unveiling crop physiological interactions. Since the 1970s, diverse mathematical models have emerged to simulate crop growth, development, and yield, finding wide applications in crop management, yield analysis, pest interactions, and climate change impact. As crop models witness an expanding array of uses, the development of targeted methods aligned with well-framed research questions becomes increasingly crucial.

Keywords: *Climate change. Crop models, Mitigation, Simulation, Yield analysis*

Introduction

Crop modeling in agriculture involves utilising quantitative data on eco-physiological processes to forecast plant growth and development, leveraging environmental conditions and management inputs. These models simulate how crops respond whether in terms of growth, yield, or other factors to a dynamic interplay of elements like management practices, water, weather, and soil parameters throughout a growing season. They serve as computational tools emulating the intricate growth patterns of crops by mathematically representing various facets within the cropping system (Nath *et al.*, 2023). The roots of crop modeling trace back to the 1960s, when researchers began integrating physical and biological principles to model agricultural systems. These models heavily rely on measurable inputs, obtained via sensors, machinery, or manual measurements, to predict desired outputs such as plant growth, crop yield, soil nitrogen levels, or crop staging.

Crop Modeling

Crop modeling involves the creation of mathematical or computational models that represent the growth, development, and yield of crops. These models are based on scientific principles and empirical data, aiming to simulate and predict various aspects of crop behaviour under specific conditions. The primary purpose of crop modeling is to provide a theoretical framework for understanding and predicting crop responses to environmental factors, management practices, and genetic variations.

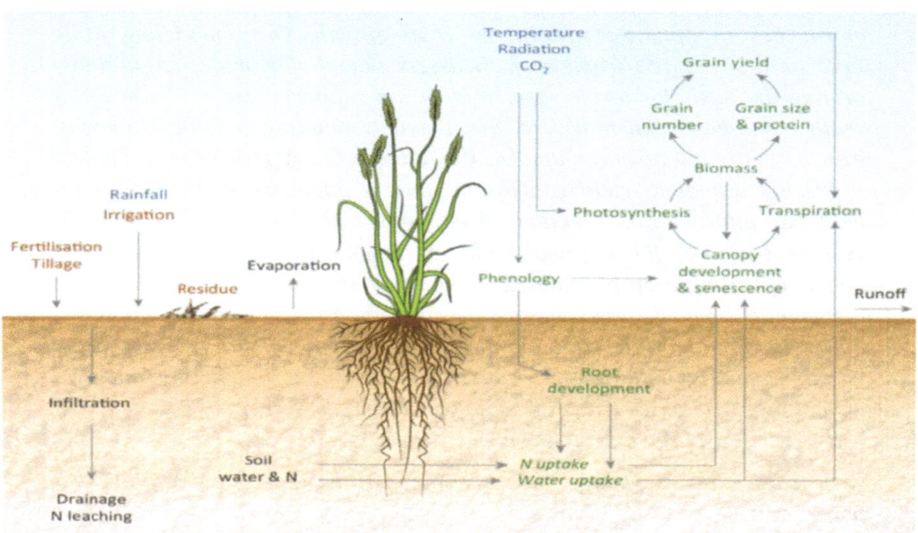

Fig. 1: Diagram illustrating the essential stages in crop growth within a crop model

Crop models integrate various parameters related to soil, climate, crop genetics, and management practices to simulate how these factors interact and influence crop growth and yield. They typically use equations and algorithms to represent biological processes. Crop models are utilised for a range of purposes, including assessing the impact of climate change on crops, optimising resource use (water, nutrients), evaluating different agricultural practices, and guiding decision-making for farmers, researchers, and policymakers.

Crop Simulation

Crop simulation involves using existing crop models or software tools to simulate specific scenarios or conditions. It is a practical application of crop modeling principles to replicate how crops might behave under given circumstances. The primary purpose of crop simulation is to replicate or mimic real-world crop behaviour under controlled or hypothetical conditions. This allows for scenario testing and analysis without conducting physical experiments.

Crop simulation utilises established crop models within software or computational tools. Researchers input specific parameters related to soil, climate, cultivars, and management practices to simulate how crops would grow and yield under those specified conditions. Crop simulation is employed for scenario testing, such as assessing the impact of changing climate variables or altering management practices on crop growth, yield, and other agronomic parameters. It aids in decision-making by providing insights into potential outcomes.

Crop modeling involves the creation and development of theoretical models representing crop growth and behaviour. Crop simulation, on the other hand, involves the practical application of these models to simulate specific scenarios or conditions. While crop modeling focuses on creating theoretical frameworks to understand crop behaviour, crop simulation applies these models practically to simulate and analyze specific scenarios or conditions. In essence, crop modeling lays the theoretical foundation for understanding crop behaviour, while crop simulation puts this theory into practice to replicate and analyse specific situations or scenarios in a controlled computational environment (Sow *et al.*, 2020).

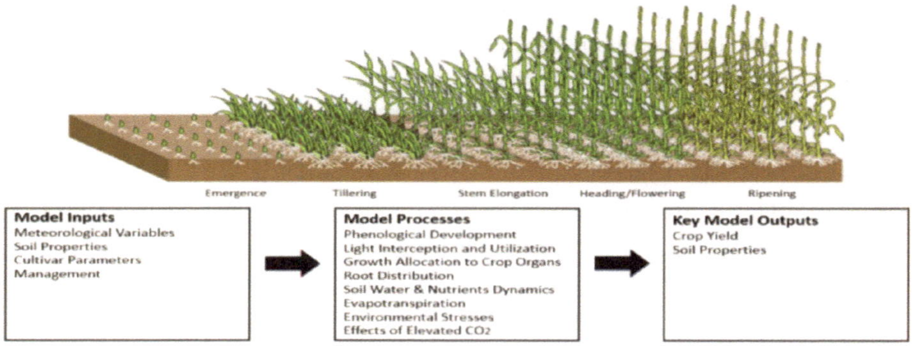

Fig. 2: Crop simulation model

The data integrated into these models encompass a wide array of factors including daily weather information such as solar radiation, temperature fluctuations, and rainfall, as well as soil attributes, initial soil conditions, cultivar specifics, and crop management practices. Remarkably, crop models can simulate multiple seasons, diverse locations, various treatments, and numerous scenarios in mere minutes.

The significance of crop models in agriculture is multifaceted. They elucidate the intricate dynamics between the atmosphere, crop, and soil, offering invaluable insights for crop agronomy, pest management, breeding, and the judicious utilisation of natural resources. Furthermore, they serve as a crucial tool to evaluate and predict the impact of climate change on crop growth and agricultural productivity (Murthy, 2004).

Crop Improvement

Crop models play a crucial role in guiding breeding programmes and envisioning ideal crop types (ideotypes). These models aid in predicting trait phenotypes and guiding selection protocols for breeding programmes aimed at achieving these ideotypes. To integrate crop models effectively into breeding, understanding the inheritance of model parameters is essential. For instance, in breeding programmes, optimising carbon allocation among plant components (like leaves, stems, rhizomes, and roots) requires both phenotypic and genotypic data along with a crop model capable of capturing the impact of various carbon allocation strategies on growth and biomass production.

Manipulating flowering time can significantly impact carbon allocation. By reducing the energy invested in reproductive structures, more biomass becomes available for harvest, aiding in the development of cultivars suited to specific regions. Adjusting carbon allocation schemes can reduce leaf area by increasing stem number or thickness while maintaining an optimum leaf area index, which improves water use efficiency. This is crucial in dry environments for biomass production.

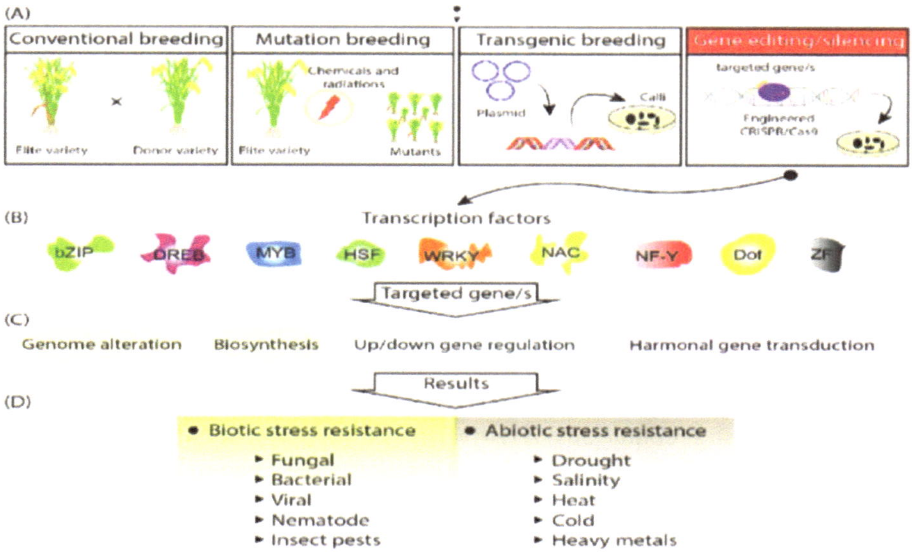

Fig. 3: Crop improvement techniques

However, altering flowering time can affect nutrient translocation to below-ground storage in perennial crops. Delayed flowering may reduce nutrient use efficiency, impacting sustainability and necessitating additional fertilisers. Bridging the gap between potential adaptation strategies and practical realisation requires locally relevant and climate-informed approaches, considering the dynamics of agricultural systems and simulating climate-crop interactions (Asseng and Cammarao, 2014).

Simulations from crop models indicate potential mean yield changes with adaptation strategies compared to non-adaptation baselines due to climate change. To enhance these approaches, it's important to:

1. Engage stakeholders to identify key vulnerabilities and timing of risks in food systems.
2. Use diverse methods that critically evaluate climate model output.
3. Increase transparency and comparability in risk assessments.

This integration of crop models into breeding and adaptation strategies contributes significantly to understanding the impacts of climate change on the global food system and socio-economic development through long-term predictions.

Crop Modeling and Simulation for Climate Smart Agriculture

Crop modeling for climate-smart agriculture is a crucial tool in addressing the challenges posed by changing climate conditions. Here's a detailed elaboration on how these models contribute to adaptation strategies, resource optimisation, and mitigation measures:

1. Adaptation Strategies

i. Understanding Crop Responses: Crop models simulate how different crops respond to changes in climatic variables such as temperature, rainfall patterns, and CO_2 levels. They forecast potential shifts in growth stages, yield, and overall productivity under different climate scenarios.

ii. Identifying Resilient Varieties: These models help in identifying and developing crop varieties that are better equipped to withstand or thrive in changing conditions. By predicting the performance of various cultivars, farmers can choose those with greater resilience to anticipated climate changes.

iii. Adaptive Management Practices: Crop models guide the implementation of adaptive management practices. Farmers can adjust planting dates, irrigation schedules, and other agronomic practices based on model predictions to optimise crop performance in evolving climates.

2. Resource Optimisation

i. Water and Nutrient Management: Models assist in optimising water and nutrient usage by predicting crop water requirements and nutrient uptake under different climate scenarios. This information helps farmers apply resources more efficiently, reducing waste and environmental impact.

ii. Precision Agriculture Techniques: These models support precision agriculture by providing insights into where and when resources should be applied. Precision techniques like variable rate irrigation or fertilisation align resources with crop needs, enhancing productivity while minimising inputs.

3. Mitigation Measures

i. Reducing Emissions: Crop models inform strategies to reduce agricultural emissions. By optimising land use practices, such as reduced tillage or cover cropping, these models help minimise greenhouse gas emissions from agricultural activities.

ii. Enhancing Carbon Sequestration: They aid in identifying cropping systems that enhance carbon sequestration in soils or biomass, contributing to carbon offset initiatives and mitigating the overall impact of agriculture on climate change (Ranjan *et al.*, 2022).

iii. Land Use Efficiency: Models assist in evaluating land use scenarios to maximise productivity while preserving natural ecosystems. Efficient land use strategies prevent deforestation or conversion of natural habitats for agriculture, thereby reducing environmental degradation.

Crop modeling for climate-smart agriculture integrates scientific understanding of crop responses to environmental changes with actionable strategies for farmers. By offering insights into adaptation, resource management, and mitigation practices, these models play a pivotal role in building resilient agricultural systems that can thrive in a changing climate while minimising environmental impact.

Components of Crop Modeling and Simulation

Each component of crop modeling for climate-smart agriculture plays a critical role in understanding and predicting how crops will respond to changing environmental conditions:

1. Climate Data Integration

i. Historical and Projected Data: Crop models rely on historical climate data to establish baseline conditions. When combined with projected data, they forecast potential future climate scenarios. This integration helps simulate how crops might behave under varying temperature, precipitation, and CO_2 levels.

ii. Scenario Analysis: By analysing different climate scenarios, models can predict how crop growth, development, and yield might change. This information assists in planning for potential climate-related challenges and adapting farming practices accordingly (Heaton *et al.*, 2010).

2. Genotype and Phenotype Integration

i. Genetic Information: Incorporating genetic data into models allows for the simulation of various crop varieties or hybrids. It predicts how these different genotypes might respond to altered climatic conditions.

ii. Phenotypic Traits: Models consider phenotypic traits (e.g., plant height, flowering time, disease resistance) to predict how specific characteristics of different crop varieties will fare under changing climates. This integration helps identify traits that contribute to resilience or better performance under certain environmental stresses.

3. Carbon and Nutrient Dynamics

i. Resource Allocation: Understanding how crops allocate resources, such as carbon and nutrients, among various plant parts (leaves, stems, roots) is crucial. Models simulate these dynamics under different climate variations to optimise resource use efficiency.

ii. Environmental Sustainability: Predicting changes in resource allocation aids in ensuring sustainable production practices. It allows for the optimisation of crop growth while minimising resource wastage or environmental impact, contributing to more sustainable agricultural systems.

4. Impact on Yield and Growth

i. Yield Forecasting: Crop models predict how changes in climate variables will impact crop yield and growth stages. This information is invaluable for farmers in making decisions regarding planting schedules, irrigation strategies, and crop selection.

ii. Adaptation Strategies: By providing insights into the expected changes in crop performance, models assist in adapting agricultural practices. Farmers can adjust their approaches, such as switching to more drought-tolerant crops or altering harvest times, to optimise productivity in changing climates.

These components collectively empower farmers, researchers and policymakers to make informed decisions about crop management, breeding programmes, and policy interventions. By harnessing the power of data integration, genotype-phenotype understanding, resource dynamics, and yield impact assessment, crop modeling contributes significantly to the development of climate-resilient agricultural systems.

Advantages and Applications of Crop Modeling and Simulation

The advantages and applications of crop modeling in climate-smart agriculture are numerous. They extend to risk assessment, scenario testing, and informing policy decisions:

1. Risk Assessment

i. Understanding Vulnerabilities: Crop models facilitate the identification and understanding of vulnerabilities in agricultural systems due to climate variability. They predict how variations in temperature, precipitation, or extreme weather events might impact crop yields and growth.

ii. Planning for Disruptions: By assessing potential risks associated with climate variability, farmers and policymakers can proactively plan for potential disruptions in crop production. This includes preparing for scenarios such as droughts, floods, or heat waves by implementing adaptive strategies in advance.

iii. Insurance and Risk Management: Crop insurance and risk management programmes benefit from crop models as they provide insights into the likelihood and severity of crop losses due to climate-related events. This information aids in designing effective insurance policies for farmers.

2. Scenario Testing

i. Virtual Experimentation: Crop models allow farmers to virtually test various scenarios before implementation. For example, they can simulate changes in irrigation patterns, altered crop varieties, or different planting schedules to assess their potential outcomes under specific climatic conditions.

ii. Optimising Strategies: Through simulated scenario testing, farmers can identify the most suitable strategies for their specific conditions. This assists in optimising resource allocation, selecting resilient crop varieties, or adjusting management practices based on anticipated climate changes.

3. Informing Policy Decisions

i. Policy Formulation: Governments and organisations use crop models to develop evidence-based policies that support climate-resilient agriculture and food security. These models provide valuable insights into potential future scenarios, enabling policymakers to make informed decisions.

ii. Resource Allocation: Policy decisions related to resource allocation, research funding, infrastructure development, and agricultural subsidies can be informed by crop models. This ensures that investments are targeted towards practices and technologies that enhance agricultural resilience in changing climates.

iii. International Collaboration: Crop models facilitate international collaboration and knowledge-sharing by providing a common platform for assessing the impacts of climate on agriculture. They help in developing coordinated strategies for global food security in the face of climate change.

In essence, crop modeling in climate-smart agriculture empowers stakeholders at various levels by offering insights into potential risks, enabling informed decision-making through scenario testing, and guiding the formulation of policies that promote sustainable and resilient agricultural systems.

Challenges and Future Strategies

The challenges and future directions in crop modeling for climate-smart agriculture are crucial for advancing its effectiveness and widespread adoption:

Challenges

i. Data Accuracy and Availability: Obtaining accurate historical data and reliable future projections remains challenging, especially in regions with limited monitoring infrastructure. Comprehensive genetic and soil data are essential for precise modeling. However, obtaining detailed genetic information about crop varieties and soil characteristics globally can be difficult.

ii. Model Validation: Continuous validation and improvement of models are essential. Ensuring that models accurately represent real-world conditions across diverse environments, crops, and management practices is a perpetual challenge.

iii. Local Adaptation: Tailoring models to local conditions is crucial for their practical application. This includes accounting for regional variations in climate, soil types, farming practices, and crop preferences. Ensuring accessibility and usability for farmers in different regions is essential for their adoption.

Future Strategies

i. Enhanced Data Accessibility and Quality: Efforts should be focused on improving data collection systems and making data more accessible. Collaborative initiatives can help enhance data quality and availability, especially in regions with limited resources.

ii. Model Improvement and Validation: Continued refinement of models is essential. Integrating new scientific findings and methodologies into models while validating them against real-world data under diverse environmental conditions enhances their reliability.

iii. Localised and User-Friendly Models: Developing models that are user-friendly and tailored to local conditions is critical. This involves engaging farmers and stakeholders in model development, ensuring they are intuitive and valuable for on-the-ground decision-making.

iv. Integrated Technology and Knowledge Sharing: Integration of technology (such as remote sensing, machine learning) into modeling frameworks can improve accuracy and efficiency. Encouraging knowledge sharing and collaboration among researchers, policymakers, and farmers promotes the development and adoption of effective strategies.
v. Capacity Building and Education: Empowering farmers and stakeholders with the knowledge and skills to use these models effectively is essential. Training programmes and educational initiatives can enhance understanding and utilization.
vi. Policy Support and Investment: Governments and organisations should support research, infrastructure, and policies that promote the development, validation, and adoption of crop models for climate-smart agriculture.

Overcoming these challenges and advancing along these future directions is crucial for the continued evolution and effectiveness of crop modeling in climate-smart agriculture. By overcoming data constraints, enhancing model precision, ensuring regional applicability, and promoting collaboration, crop models can effectively function as a potent instrument for steering sustainable agricultural practices. This empowerment equips farmers to adapt and flourish amidst the challenges posed by a shifting climate. Adaptation involves aligning with the present and anticipated impacts of climate change, while mitigation entails mitigating the severity of these impacts by averting or diminishing the release of greenhouse gases (GHG) into the atmosphere. The adaptation strategies include the application of organic fertilizers, changing planting dates and growing short-duration crop varieties. The mitigation actions involve planning and zoning, floodplain protection, property acquisition and relocation, or public outreach projects. Illustrations of preparedness measures include the implementation of disaster warning systems, procurement of radio communications equipment, and the execution of emergency response training.

Conclusion

In conclusion, crop modeling serves as the foundational bedrock for understanding and predicting crop behaviour. It involves the intricate creation of mathematical or computational models rooted in scientific principles and empirical data. These models encompass various parameters, such as soil characteristics, climate factors, genetics, and management practices, aiming to simulate and forecast crop growth, development, and yield under specific conditions. On the other hand, crop simulation stands as the practical application of these meticulously crafted models. It entails employing existing crop models or software tools to virtually replicate how crops would respond under specified scenarios or conditions. By inputting precise parameters related to soil, climate, cultivars, and agricultural practices, researchers utilise crop simulation to mimic real-world crop behaviour, enabling scenario testing and analysis without physical experimentation.

Ultimately, while crop modeling establishes the theoretical framework for comprehending crop behaviour, crop simulation translates this theory into practical utility, enabling the replication and analysis of specific scenarios within a controlled computational environment. Together, they form an integral part of agricultural research and decision-making, empowering stakeholders with predictive insights and informed strategies for sustainable and efficient crop management.

References

Asseng, S., and Cammarao, D. (2014). Simulation modeling: Application in cropping system. Encyclopedia of Agriculture and Food systems.

Heaton, E.A., Dohleman, F.G., Miguez, F.E., *et al.* (2010). Modeling for crop improvement. Advances in Botanical Research, 56.

Murthy, R.K. (2004). Crop growth modelling and its application in agricultural meteorology. Satellite Remote Sensing and GIS Applications in Agricultural Meteorology, pp. 235-261

Nath, D., Suman, S.N., Raigar, B. L., Sow, S., and Ranjan, S.R. (2023). Carbon Dynamics in Soil in Relation to Climate Change. In: Climate Smart Approaches towards Sustainable Crop Production, Vital Biotech Publication, pp. 89-118.

Ranjan, S., Sow, S., Kumari, J., Kumar, N., Roy, D.K., and Pramanick, B. (2022). Use of Remote Sensing in Agriculture. Agriculture Letters, 3(9), 48-52.

Sow, S., Ranjan, S., Ghosh, M., Kumar, S., and Choudhury, S.R. (2020). Crop Residues as Potential Sustainable Source of Silica in the Era of Climate Change: A Review. International Journal of Current Microbiology and Applied Sciences, 10(11),1629-1641.

4

Government Initiatives for Sustainable Agriculture in India

Anil Chavan, Pundlik Waghmare
Sharad Chenalwad and Santosh Shinde

Department of Agronomy, Vasantrao Naik Marathwada Krishi Vidyapeeth Parbhani (MS), India

Abstract

Agriculture is the backbone of the Indian economy, with over 50% of the population dependent on it for their livelihood. In recent years, the adoption of advanced technologies has played a crucial role in the prosperity of Indian farmers, contributing significantly to economic growth. Precision farming stands out as a key approach, leveraging modern equipment and technology to enhance crop yields while minimising environmental impact. The implementation of state-of-the-art agricultural tools empowers farmers to achieve higher productivity with more efficient resource utilisation. Lesser need for water, soil optimisation elements such as fertilisers and fewer workforces lead to cost-effectiveness and higher profits. The agricultural sector in India has implemented various initiatives targeting climate change challenges. Under different schemes like RKVY, BGREI, NFSM, and NMSA, programs like crop diversification, rainfed area development, and organic farming (PKVY) have been employed to bolster resilience. These efforts aim to reduce carbon emissions by encouraging sustainable practices like agroforestry and soil health improvement (SHC), aligning with the broader goal of making agriculture more resilient, adaptive, and eco-friendly. These initiatives collectively strive to enhance agricultural productivity, mitigate climate impacts, ensure food security, and promote environmental sustainability while supporting farmer livelihoods.

Keywords: *Government Initiatives, NFSM, NMSA, NAPCC, RKVY, Sustainable Agriculture*

Introduction

Sustainable agriculture focuses on practices that ensure the long-term well-being of the environment, society, and economy. In India, the government has instituted various policies and initiatives to encourage sustainable farming practices and enhance farmers' income. Embracing sustainability is crucial for making agriculture both profitable and environmentally friendly for future generations, and technology emerges as the key enabler for achieving this goal. Technological advancements empower farmers to make informed decisions regarding optimal planting and harvesting times, as well as the judicious use of water and fertilisers. This leads to reduced waste, increased crop yields, and ultimately higher profits for farmers. Additionally, technology plays a significant role in mechanisation, encompassing the use of modern agricultural implements that multitask while minimising soil impact (McFadden *et al.*, 2023). Addressing major challenges such as crop residue management through environmentally friendly technological methods not only enhances efficiency and productivity but also reduces the physical labor required, transforming farming into a less strenuous occupation.

Government Initiatives for Farmers

In addition to technology, various government initiatives are in place to support Indian farmers. The Pradhan Mantri Fasal Bima Yojana (PMFBY), for instance, is a crop insurance scheme designed to offer financial assistance to farmers in the event of crop loss caused by natural disasters or unforeseen circumstances. This initiative aims to mitigate the financial risks associated with farming and provide a safety net for farmers.

Another noteworthy initiative is the Pradhan Mantri Kisan Samman Nidhi (PM-KISAN), which extends financial assistance to small and marginal farmers to facilitate the purchase of agricultural inputs like seeds and fertilisers. The objective is to enhance the income of small farmers and elevate their overall standard of living.

Furthermore, the Indian government has launched several schemes to promote the adoption of modern farming implements and technology. For example, the National Food Security Mission (NFSM) offers financial assistance to farmers for acquiring modern agricultural tools, intending to encourage mechanisation and boost productivity in the agricultural sector. India has implemented various government initiatives aimed at promoting sustainable agriculture practices across the country. Some key initiatives include:

1. National Mission for Sustainable Agriculture (NMSA)

Launched in 2010, this mission aims to promote sustainable agriculture through various strategies like soil health management, water use efficiency, and promoting climate-resilient agricultural practices. The National Mission for Sustainable Agriculture (NMSA) is a comprehensive initiative launched by the Government of India in 2010. It aims to promote sustainable agriculture practices and address various challenges faced by the agricultural sector, including climate change, soil degradation, water scarcity, and declining farm productivity.

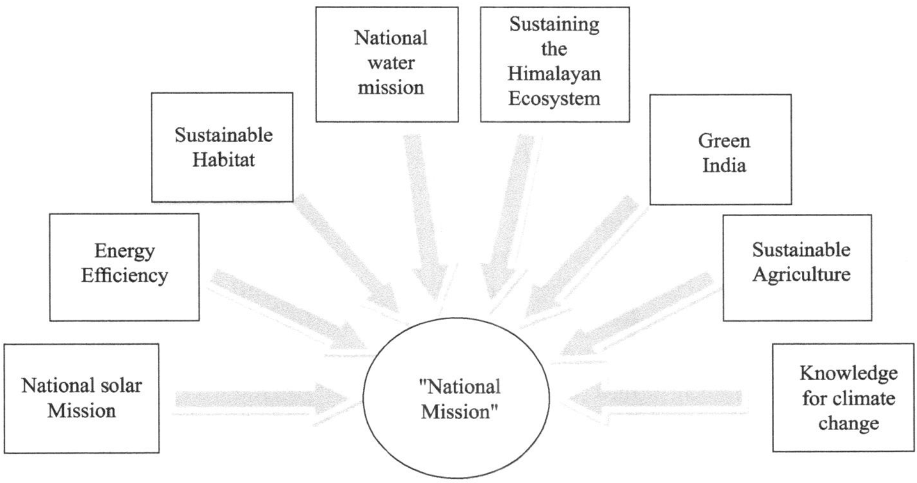

Fig. 1: Indian Government national mission for sustainable agriculture development

The mission focuses on enhancing agricultural productivity in an environmentally sustainable manner.

Objectives of NMSA

i. Promotion of Climate-resilient Farming: NMSA aims to encourage farmers to adopt climate-resilient agricultural practices suited to their specific agro-climatic zones. This involves promoting crop diversification, use of stress-tolerant crop varieties, and conservation agriculture techniques.

ii. Soil Health Management: One of the primary objectives is to improve soil health through balanced nutrient management, minimising soil erosion, and promoting organic farming practices. The Soil Health Card Scheme is a part of NMSA, providing farmers with information about their soil's health status and necessary recommendations for improving soil fertility.

iii. Water Use Efficiency: The mission emphasises the efficient use of water resources in agriculture. It encourages the adoption of water-saving irrigation techniques such as drip and sprinkler irrigation, along with proper water management practices.

iv. Promotion of Integrated Farming Systems: NMSA supports the adoption of integrated farming systems that combine crop cultivation, animal husbandry, fisheries, and agroforestry. This helps in maximising resource utilisation and increasing farmers' income.

v. Capacity Building and Awareness: NMSA focuses on educating and training farmers about modern agricultural practices, sustainable technologies, and best management practices to improve productivity and sustainability (Jeppesen *et al.*, 2016).

Rainfed Area Development (RAD) focuses on improving productivity and livelihoods in rainfed areas by promoting watershed management, soil moisture conservation, and crop diversification. Soil Health Management: This component emphasises soil testing, the distribution of soil health cards, and promoting organic farming practices to improve soil health and fertility. NMSA supports initiatives aimed at enhancing livelihoods of farmers by promoting sustainable agricultural practices and providing alternative sources of income through allied activities like animal husbandry, fisheries, and agroforestry. The mission also addresses risk management in agriculture by promoting crop insurance schemes and encouraging the adoption of climate-resilient crops (Zhang, 2016).

Overall, NMSA is a holistic approach toward sustainable agriculture, aiming to ensure food security, improve farmers' livelihoods, and conserve natural resources for future generations. It integrates various components and strategies to achieve the goal of sustainable agricultural development in India.

2. The Pradhan Mantri Krishi Sinchayee Yojana (PMKSY)

It is a flagship scheme launched by the Government of India in 2015 with the primary objective of improving farm productivity and ensuring efficient water utilisation in agriculture. The scheme aims to expand irrigation coverage and enhance water use efficiency "Per Drop More Crop."

Objectives of PMKSY

i. Expansion of Irrigation Coverage: PMKSY aims to increase the irrigated area in the country by providing better access to water for agricultural purposes. It focuses on developing and expanding irrigation facilities through various means, including traditional surface irrigation, micro-irrigation systems, and rainwater harvesting.

ii. Water Use Efficiency: The scheme emphasises the efficient use of water resources in agriculture. It promotes the adoption of modern irrigation techniques such as drip irrigation, sprinkler irrigation, and precision irrigation to ensure optimal utilisation of water resources (Sow *et al.*, 2022).

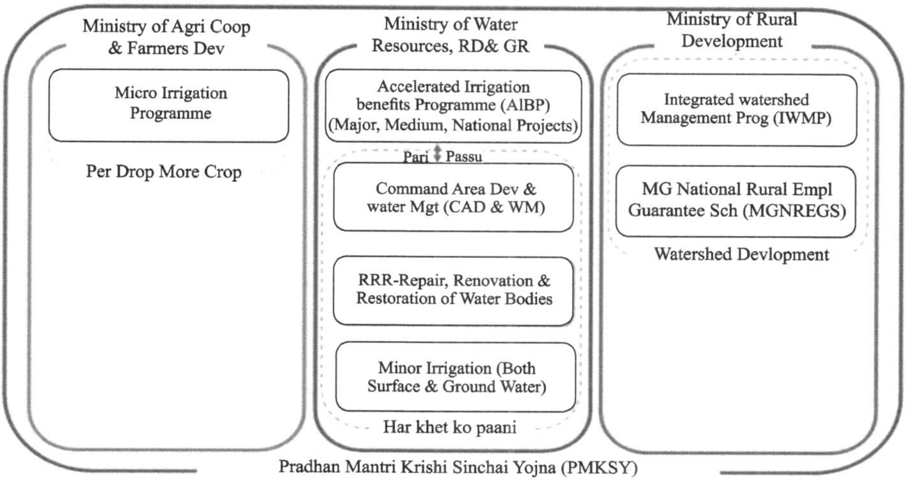

Fig. 2: Pradhan Mantri Krishi Sinchai Yojana (PMKSY)

iii. Improving On-Farm Water Management: PMKSY aims to encourage farmers to adopt better on-farm water management practices. This includes promoting measures like lining canals, constructing water storage structures, and watershed development to conserve water and reduce wastage.

iv. Promotion of Sustainable Agriculture: Alongside water management, the scheme promotes sustainable agricultural practices that are aligned with efficient water usage. This involves encouraging crop diversification, promoting organic farming, and providing training and support to farmers for adopting modern agricultural techniques.

Components of PMKSY

i. Accelerated Irrigation Benefit Programme (AIBP): This component focuses on completing irrigation projects and creating new infrastructure for irrigation to bring more areas under assured water supply for agriculture.

ii. Har Khet Ko Pani (HKKP): The scheme aims to improve water use efficiency through measures like micro-irrigation, surface water management, and rainwater harvesting. It promotes the concept of 'More Crop Per Drop' by ensuring water reaches every field.

iii. Per Drop More Crop (PDMC): PDMC emphasises efficient water management and encourages farmers to adopt precise irrigation technologies, drip and sprinkler irrigation systems, and other modern practices to reduce water consumption while increasing crop productivity.

iv. Watershed Development: PMKSY supports watershed management and development to conserve soil and water resources. It aims to create sustainable models for rainwater harvesting and groundwater recharge.

PMKSY plays a crucial role in enhancing agricultural productivity, reducing dependency on monsoons, and promoting sustainable water management practices in agriculture across the country. It aims to ensure water security for farmers and improve their livelihoods by enhancing agricultural productivity

3. The Paramparagat Krishi Vikas Yojana (PKVY)

It is a scheme launched by the Indian government under the Ministry of Agriculture and Farmers Welfare. It aims to promote organic farming among Indian farmers and encourage them to adopt eco-friendly practices. Here's a detailed overview of PKVY:

Objectives

i. Promotion of Organic Farming: The primary goal is to promote chemical-free organic farming practices among farmers.

ii. Sustainability: Encouraging sustainable agricultural practices that ensure soil health, biodiversity conservation, and environmental preservation.

iii. Enhanced Income: Facilitating higher incomes for farmers by promoting organic produce, which generally commands better prices in the market.

Key Features

i. Group Approach: PKVY operates on a group-based model, encouraging farmers to form clusters or groups of 50 farmers each. These groups collectively take up organic farming practices in their fields.

ii. Financial Support: The scheme provides financial assistance to farmers to help them transition from conventional farming to organic methods. This includes support for inputs like organic manure, bio-fertilisers, bio-pesticides, vermicompost, etc.

iii. Training and Capacity Building: Farmers receive training on organic farming techniques, methods of organic inputs preparation, composting, seed treatment, and other aspects of organic cultivation. This helps in enhancing their knowledge and skills in sustainable farming.

iv. Certification: The produce from PKVY-supported fields can be certified as organic. Certification adds value to the produce and expands market access due to the growing demand for organic products.

v. Incentives: Incentives are provided to farmers and groups for following organic practices. These can be in the form of awards, recognition, or additional financial support based on their performance in adopting and sustaining organic farming.

vi. Monitoring and Evaluation: Regular monitoring and evaluation are conducted to ensure the proper implementation of the scheme and to measure its impact on soil health, crop yield, and farmer income.

Benefits

i. Environmental Benefits: Reduced chemical usage leads to improved soil health, water conservation, and biodiversity preservation.

ii. Health Benefits: Organic produce is free from harmful chemical residues, making it healthier for consumers.

iii. Economic Benefits: Farmers can fetch higher prices for their organic produce, leading to increased incomes.

PKVY plays a crucial role in promoting sustainable agriculture and encouraging farmers to adopt organic practices. By providing financial support, training, and incentives, the scheme aims to transform conventional farming into a more ecologically friendly and economically viable model for farmers in India.

4. The Soil Health Card Scheme (SHC)

It is an initiative launched by the Government of India to assess the nutrient status of agricultural soils and provide farmers with personalised recommendations to improve soil health and enhance crop productivity. Here's a detailed look at the scheme:

Objectives

i. Assessment of Soil Health: The primary goal is to evaluate the nutrient levels and health status of agricultural soils across the country.

ii. Farmer Empowerment: Empowering farmers with information about the specific requirements of their soil and crops, thereby aiding them in making informed decisions regarding fertiliser use and crop selection.

iii. Sustainable Agriculture: Promoting sustainable agricultural practices by encouraging balanced and judicious use of fertilisers to improve soil fertility and productivity.

Key Features

i. Soil Testing: Under the Soil Health Card Scheme, soil samples are collected from farmers' fields and tested at government-certified soil testing labs. These tests analyse various parameters such as pH levels, macro and micronutrient content and organic carbon levels.

ii. Customised Recommendations: Based on the soil test results, farmers receive personalised Soil Health Cards. These cards contain information about the current nutrient status of their soil and provide recommendations on the types and quantities of fertilisers needed for optimal crop growth.
iii. Periodic Updates: The scheme aims to provide Soil Health Cards to farmers once every two years for irrigated areas and once every three years for non-irrigated areas. This periodicity ensures that farmers have updated and relevant information about their soil health.
iv. Awareness and Training: Farmers are educated and trained on various aspects related to soil health management, including the importance of balanced nutrient application, soil conservation and organic farming practices.
v. Technology Integration: The scheme utilises technology for efficient data collection, analysis, and dissemination of information. Online portals and mobile applications have been developed to facilitate easy access to soil health-related information (Anon, 2017).

Benefits

i. Optimised Fertilizer Use: Farmers receive precise recommendations, reducing unnecessary use of fertilisers and minimising the risk of overuse, which can lead to soil degradation and environmental pollution.
ii. Increased Productivity: By adopting soil health-based recommendations, farmers can improve the productivity and quality of their crops.
iii. Cost Savings: Efficient use of fertilisers based on soil health cards can result in cost savings for farmers while maintaining or enhancing crop yields.

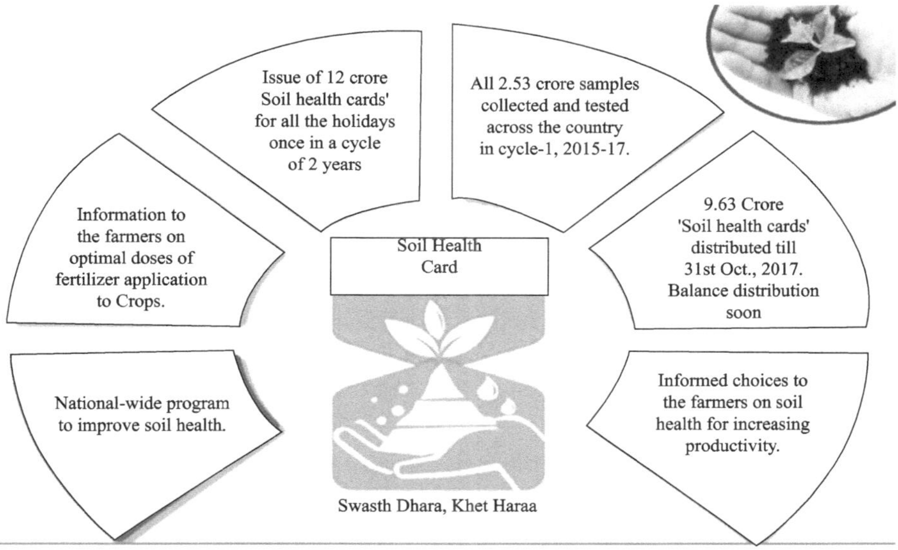

Fig. 3: Soil Health Card

The Soil Health Card Scheme is an essential step towards promoting sustainable agriculture by focusing on soil health and providing farmers with actionable information. By empowering farmers with personalised recommendations, the scheme aims to enhance agricultural productivity, reduce input costs, and contribute to the overall welfare of farmers and the environment.

5. The Rashtriya Krishi Vikas Yojana (RKVY)

It is a comprehensive agricultural development scheme launched by the Government of India to accelerate the growth of the agriculture sector. It was introduced in 2007 and underwent several phases and modifications to address the diverse needs of agriculture across states and regions. Here's a detailed overview:

Objectives

i. Promotion of Agricultural Growth: The primary aim is to foster holistic development in agriculture by enhancing productivity, ensuring a remunerative return to farmers, and augmenting overall agricultural growth.

ii. Strengthening Infrastructure: Improving agricultural infrastructure, including irrigation, storage facilities, roads, markets, and agro-processing units, to support efficient farming practices and value addition (Nicoleta *et al.*, 2019).

iii. Sustainable Development: Encouraging sustainable practices, technology adoption, and diversification to mitigate risks associated with climate change and promote eco-friendly farming methods.

Key Features

i. Flexible Funding: RKVY provides flexibility to states in planning and implementing agriculture-related projects. It allocates funds to states based on their requirements and proposals submitted for various agricultural activities.

ii. Area of Focus: The scheme covers a wide spectrum of agricultural activities, including crop diversification, horticulture, animal husbandry, fisheries, watershed management, agricultural research, and extension services.

iii. Integrated Approach: Emphasis is placed on an integrated approach to agriculture, linking various components such as production, processing, marketing, and infrastructure development to ensure a comprehensive boost to the sector.

iv. Technology Adoption: Encouraging the adoption of modern agricultural practices, high-yielding varieties, improved seeds, efficient irrigation methods, and mechanisation to enhance productivity and reduce post-harvest losses.

v. Capacity Building: Providing training, education, and extension services to farmers to improve their knowledge, skills, and awareness about modern agricultural practices and technologies.

vi. Monitoring and Evaluation: Regular monitoring and evaluation mechanisms are in place to ensure the effective utilisation of funds and the achievement of set targets and objectives.

Benefits

i. Increased Productivity: The scheme aims to increase agricultural productivity and crop yields through the adoption of modern technologies and better farming practices.

ii. Diversification: Encouraging farmers to diversify their crops and adopt allied activities like horticulture, animal husbandry, and fisheries to reduce risks and enhance income.

iii. Infrastructure Development: Strengthening agricultural infrastructure leads to improved connectivity, storage facilities, and market access, benefiting farmers and reducing post-harvest losses.

RKVY plays a pivotal role in driving agricultural growth and development across states in India. By providing financial support, encouraging technology adoption, and promoting sustainable practices, the scheme aims to enhance the livelihoods of farmers, ensure food security, and contribute to the overall growth of the agriculture sector in the country.

6. The Mission for Integrated Development of Horticulture (MIDH)

It is a centrally sponsored scheme launched by the Government of India to promote holistic growth and development of the horticulture sector. Introduced in 2014-15, MIDH aims to enhance horticultural production, improve productivity, and ensure the integrated development of the horticulture industry. Here's a detailed overview of MIDH:

Objectives

i. Promotion of Horticultural Production: The primary objective is to increase horticultural production and productivity by adopting modern technologies, best practices, and scientific interventions.

ii. Integrated Development: Encouraging an integrated approach that covers the entire horticulture value chain, including production, post-harvest management, processing, marketing, and export of horticultural produce.

iii. Sustainable Practices: Promoting sustainable horticulture practices by focusing on efficient water management, soil health improvement, use of organic inputs, and minimising post-harvest losses.

iv. Capacity Building: Providing training, education, and skill development programs for farmers, stakeholders, and entrepreneurs involved in the horticulture sector.

Key Features

1. Area of Focus: MIDH covers various components of horticulture, including fruits, vegetables, flowers, aromatic plants, spices, and medicinal plants.
2. Financial Assistance: The scheme provides financial support to states, union territories, and other implementing agencies for the development of horticulture infrastructure, adoption of new technologies, and capacity building initiatives.
3. Sub-Schemes: MIDH comprises several sub-schemes, including the National Horticulture Mission (NHM), Horticulture Mission for North East & Himalayan States (HMNEH), National Mission on Medicinal Plants (NMMP), and National Bamboo Mission (NBM).

4. Support for Farmers: Farmers receive assistance for setting up orchards, establishing nurseries, adopting protected cultivation techniques, promoting organic farming, and adopting precision farming methods.
5. Market Linkages: Emphasis is placed on creating market linkages and establishing farmer-producer organisations (FPOs) to ensure better access to markets and fair prices for horticultural produce.
6. Research and Development: Encouraging research and development activities to introduce new varieties, improve cultivation practices, and develop innovative technologies for the horticulture sector.

Benefits

1. Income Generation: MIDH aims to enhance the income of farmers by promoting high-value horticultural crops and providing support for improved production practices.
2. Employment Opportunities: The development of the horticulture sector generates employment opportunities, especially in rural areas, across various stages of the value chain.
3. Diversification and Nutrition: Encouraging diversification of crops leads to improved nutrition and dietary diversity, benefiting both producers and consumers.

MIDH plays a crucial role in fostering the growth and development of the horticulture sector in India. By focusing on integrated development, technology adoption, capacity building, and market linkages, the scheme aims to improve the livelihoods of farmers, ensure food security, and contribute to the overall growth of the agricultural economy.

7. Zero Budget Natural Farming (ZBNF)

It is an agricultural practice that promotes chemical-free farming techniques without relying on external inputs or investment. Developed by agriculturist Subhash Palekar in India, ZBNF emphasises natural and organic methods, aiming to minimise costs while increasing farmers' incomes. Here's an in-depth look at ZBNF:

Principles

1. Zero Budget Approach: ZBNF advocates for farming without using credit or spending money on purchased inputs like seeds, fertilisers, pesticides, and other external agents. It utilises natural resources available on the farm itself.
2. Natural Inputs: Instead of external inputs, ZBNF relies on natural resources like cow dung, cow urine, crop residues, mulching, and

botanical extracts to maintain soil fertility, control pests and diseases, and enhance plant growth.

Fig. 4: Different components of zero budget natural farming

3. Biodiversity and Multicropping: Encouraging the cultivation of diverse crops and promoting polyculture or mixed cropping to enhance soil health, increase productivity, and reduce the risk of crop failures.
4. Soil Health Improvement: Focuses on improving soil health through techniques like mulching, cover cropping, and using compost or cow-based manure to enhance soil structure, fertility, and water retention capacity.
5. Promotion of Indigenous Knowledge: ZBNF values traditional farming wisdom and indigenous practices, integrating them with modern scientific techniques to create sustainable and eco-friendly farming systems.

Key Practices

1. Beejamrutha and Jeevamrutha: These are microbial culture solutions prepared using local cow dung, cow urine, jaggery, gram flour, and water. Beejamrutha is used for seed treatment, while Jeevamrutha is a soil application to enhance microbial activity.
2. Mulching: Covering the soil with crop residues or mulch to retain moisture, suppress weeds, improve soil fertility, and prevent erosion.
3. Bijamrita: Seed treatment using natural extracts to protect seeds from pests and diseases and to enhance germination and plant growth.
4. Inter-cropping and Crop Rotation: Planting different crops in the same field or rotating crops to maintain soil health, control pests, and improve nutrient availability.

Benefits

1. Cost Reduction: By eliminating the need for external inputs, ZBNF reduces production costs significantly, making farming economically sustainable for small and marginal farmers.

2. Environmental Sustainability: ZBNF promotes eco-friendly practices, reducing the negative environmental impact associated with chemical-intensive farming, such as soil degradation and water pollution.
3. Improved Soil Health: Focus on natural soil fertility enhancement techniques results in healthier soils with better water retention, increased microbial activity, and improved nutrient availability.
4. Higher Farmer Incomes: Despite the zero budget approach, ZBNF has shown increased yields and better market value for organic produce, leading to higher incomes for farmers.

Challenges

1. Transition Period: Farmers transitioning from conventional farming to ZBNF may face initial challenges in adapting to new practices and experiencing temporary yield fluctuations.
2. Knowledge and Awareness: Creating awareness and providing training and support to farmers about ZBNF practices can be a challenge, especially in remote or less accessible areas.
6. Zero Budget Natural Farming presents a promising alternative to conventional agriculture, offering sustainable, low-cost farming practices that focus on environmental conservation, soil health, and farmer prosperity. However, its widespread adoption requires continuous support, education, and infrastructure development to empower farmers and ensure long-term success.

These initiatives aim to enhance agricultural productivity, ensure food security, conserve natural resources, and improve the livelihoods of farmers while promoting sustainable agricultural practices.

Conclusion

The policies and initiatives of the Indian government in sustainable agriculture aim to augment farmers' income, guarantee food security, and advocate for environmentally friendly farming practices. These efforts aim to achieve long-term agricultural sustainability while improving the livelihoods of farmers across the country. The integration of technology, particularly precision farming methods, has revolutionised Indian agriculture, bolstering both farmer prosperity and economic growth. These modern practices, encompassing advanced equipment and techniques, have significantly optimised crop yields while minimising environmental impact. Through precision farming, the judicious use of resources like water and fertilisers has not only led to cost-effectiveness but also heightened profitability for farmers. Moreover, the concerted efforts within the agricultural sector, witnessed through initiatives like RKVY, BGREI, NFSM, and NMSA under the NAPCC umbrella, exemplify India's commitment to combatting climate change challenges. The focus on crop diversification, rainfed area development, organic farming via PKVY, and other sustainable practices aims at reducing carbon emissions and fortifying agricultural resilience. Collectively, these endeavors strive towards a dual objective: augmenting agricultural productivity while mitigating the adverse impacts of climate change. They ensure food security, promote environmental sustainability, and crucially support the livelihoods of millions reliant on agriculture. As technology and sustainable practices continue to intertwine, Indian agriculture progresses toward a more adaptive, resilient, and environmentally conscious future.

References

Anon. (2017). Technology in agriculture: data-driven farming. U.S. Senate, Subcommittee on Consumer Protection, Product Safety, Insurance, and Data Security. Committee on Commerce, Science, and Transportation, Washington, DC.

Jeppesen, J. H., Jacobsen, R. H., Jørgensen, R. N., & Toftegaard, T. S. (2016). Towards Data-Driven Precision Agriculture using Open Data and Open Source Software. International Conference on Agricultural Engineering, June 26-29, 2016, Aarhus, Denmark, 01-06.

McFadden, J., Njuki, E., & Griffin, T. (2023). Precision Agriculture in the Digital Era: Recent Adoption on U.S. Farms, EIB-248, U.S. Department of Agriculture, Economic Research Service. Economic Information Bulletin Number 248, 01-52.

Nicoleta, T., Stavros, S., & Manos, R. (2019). Data-Driven Decision Making in Precision Agriculture: The Rise of Big Data in Agricultural Systems. Journal of Agricultural & Food Information, 20(4), 344-380.

Sow, S., Ranjan, S., Ghosh, M., Kumar, N., Kumar, S., & Roy, D.K. (2022). Impact of COVID-19 Pandemic on Indian Agriculture. Chemical Science Review and Letters, 1 (44), 489-494.

Zhang, Q. (2016). Precision agriculture technology for crop farming. CRC Press Taylor & Francis Group, Broken Sound Parkway NW, Suite, Boca Raton, FL-33487-2742.

5

Innovative Management Practices for Diseases and Pests in Field Crops

Shashi Kant Ekka[1], Reena Kujur[2] and Johnson Lakra[3]

[1]Department of Horticulture, SHUATS, Prayagraj-211007, U.P., India

[2]Department of Biological Sciences, SHUATS, Prayagraj-211007, U.P., India

[3]College of Agriculture & Research Station, Kunkuri-496225 C.G., India

Abstract

The agricultural sector faces increasing challenges in managing diseases and pests in field crops while minimising environmental impact and ensuring sustainable production. This chapter provides a glimpse into the dynamic landscape of innovative management practices for diseases and pests in field crops. By implementing a multidisciplinary and cooperative method, agriculture experts are poised to overcome existing challenges, ushering in a new era of productivity, sustainability, and resilience in the face of evolving agricultural threats. Innovative biological control methods, such as the use of beneficial insects and microorganisms, are gaining prominence. These approaches leverage the natural predators and antagonistic properties of certain organisms to mitigate the impact of pests and diseases, fostering a balanced ecosystem within agricultural landscapes. By adopting a holistic approach and integrating these innovative practices, farmers can foster resilient and sustainable agricultural systems that effectively manage diseases and pests, contributing to global food security and environmental conservation.

Keywords*: Biological control, Disease, Pest management, Sustainable production*

Introduction

The managemment of pests and diseases in agriculture is challenging for farmers. Antibiotics and chemical pesticides are commonly used in traditional methods, which can be harmful to human health, the environment, and the long-term sustainability of agricultural practices. This article explores innovative approaches to pest and disease management in agriculture that are changing the game (Lykogianni *et al.*, 2021). Techniques for controlling diseases and pests in field crops are necessary to keep agricultural systems healthy and productive. Farmers and agricultural professionals can prevent, monitor, and control pests and diseases that can significantly affect crop yields in a variety of ways. These administrators are practicing strategies to find a balance between restricting the use of chemical inputs, promoting sustainable agriculture, and maintaining financial viability.

Principle of Integrated Pests and Disease Management (IPDM)

- Suppression and Prevention: The elimination and prevention of pests are the first tenets of IPM. IPM aims to prevent a single pest from dominating a cropping system or causing damage, not eliminate all pests (Barzman *et al.*, 2015). Three distinct sub-principles are combined in this principle: crop rotation and management, ecology, and a multi-pest strategy in combination The papers in this Special Issue discuss each of these guiding principles. A genuine illustration of the mix of strategies and multi-pest approaches is crafted by Poggi *et al.* (2021), who discussed ways to control wireworms in field crops. A risk assessment based on the production context (such as crop, climate, soil characteristics, and landscape) and monitoring of adult and/or larval populations should be the first steps in any new agro-ecological strategy.
- Non-Chemical Approaches: For pest control, a wide range of direct, non-chemical methods are available. A few examples include soil solarisation, trap development, mechanical control, natural control, and various biotechnical strategies. However, there is a wide range in their availability, efficiency, and utility (Goldel *et al.*, 2020). Give a list of a wide range of other methods that have been used to control the largest potato pest in the world, the Colorado potato beetle (Leptinotarsa decemlineata). In addition, they compare each method to conventional insecticides and classify the benefits and drawbacks of each.
- Observing and Choice in view of Checking and Edges: Climate and agronomic circumstances in various regions can collectively influence the abundance of pests and their ability to damage the same crop. As a result, a species' life cycle and where it occurs in newly developed areas may differ from where the crop has been grown for a long time. Focusing on the science and biology of significant pest species and their natural adversaries is important to develop proper pest management techniques for crop yield (Cruces *et al.*, 2020).
- Pesticide Selections and Less Use of Pesticides: Two IPM principles directly target pesticides, recommending that the pesticides used

be as specific as possible to the target pests and have the fewest side effects on human health, non-target organisms, and the environment as possible (Barzman *et al.*, 2015). Also, decreasing the dose, recurrence of use, and turning to the incomplete utilisation of pesticides adds to the objective of IPM to diminish or limit dangers to human well-being and the environment. As a result, seed treatment has been regarded as an ecologically sound approach (Sahoo *et al.*, 2023a).

- Anti-Resistance Strategies: Since humans started using chemicals to protect plants, there have been cases of pest resistance. At the point when a nuisance becomes safe, the insect poison is utiliseutilised all the more much of the time, and in the long run, it should be supplanted as its viability winds down. Kadoić Balaško *et al.* (2021) used population genetics and geometric morphometric to try to find a reliable pattern of differences in the resistance type of the western corn rootworm.
- Evaluations: According to Barzman *et al.* (2015), several principles encourage farmers to evaluate the effectiveness of the crop protection measures they use. This is a crucial component of effective management. However, farmers often have very limited knowledge of pests and pest management strategies. Hence, many analysts emphasise the need for ongoing professional development of farmers, not only to provide managerial support but also to offer guidance on economic practices (Kuramoto *et al.*, 2002).

Methods of IPDM

The term "Integrated Pest and Disease Management" (IPDM) refers to a strategy that combines various approaches for economically and environmentally sustainable pest and disease management of field crops. Here are the critical parts of a Coordinated Nuisance and Illness Framework:

1. Cultural Practices

- Crop Rotation: Rotate crops to disrupt pest and disease life cycles.
- Polyculture/Diversification: Planting a variety of crops reduces the risk of widespread infestations.
- Use of Resistant Varieties: Select and plant crop varieties that are resistant to pests and diseases.

2. Biological Control

- Predators and Parasitoids: Introduce or conserve natural enemies of pests, such as ladybugs, spiders, or parasitic wasps.

- Microbial Control: Use beneficial microorganisms like bacteria, fungi, and viruses to control pests.

3. Mechanical and Physical Controls

- Handpicking: Physically remove pests from plants.
- Traps and Barriers: Use traps, sticky traps, or barriers to reduce pest populations.
- Mechanical Cultivation: Plow or till the soil to disrupt pest life cycles.

4. Chemical Controls

- Use of Pesticides: Use chemical pesticides judiciously and only when necessary, following proper guidelines and safety measures.
- Bio pesticides: Use natural substances, such as neem oil or insecticidal soaps, as alternatives to synthetic pesticides.

5. Monitoring and Early Detection

- Regularly monitor crops for signs of pests and diseases.
- Implement early detection systems to identify problems before they become widespread.

6. Crop Residue Management

- Properly manage crop residues to reduce the overwintering sites for pests and diseases.

7. Farm Hygiene

- Keep the farm clean by removing weed hosts, volunteer crops, and other potential breeding sites for pests (Sow and Ranjan, 2020).

8. Information and Education

- Educate farmers about the principles and practices of IPDM.
- Provide information on pest and disease identification, life cycles, and effective control measures.

9. Decision Support Systems

- Utilise technology and data to make informed decisions about pest and disease management.
- Implement weather-based models for predicting pest and disease outbreaks.

10. Community Participation

- Encourage collaboration among farmers, extension services, researchers, and other stakeholders for effective pest and disease management.

Alternatives to Conventional Farming

The need for conventional agricultural systems that utilise fewer synthetic crop protection products and fertilisers with improved efficiency is crucial. Shifting to alternative farming systems, such as those relying on ecosystem services like wildflowers in field margins, is increasingly important. These alternative systems are often referred to as low-input farming systems, as they are expected to use fewer chemicals and less machinery compared to conventional systems. However, these low-input systems must not only maintain short-term yield production but also strive for increased yields in the longer term, a concept known as sustainable intensification. Sustainable intensification aims to produce more food while delivering environmental and social benefits.

One well-known alternative system is organic agriculture, which abstains from synthetic inputs and genetically modified crops. In the European Union, the total organic area covered approximately 13.8 million hectares in 2019, constituting 8.5% of the total agricultural area. Disease management in organic agriculture focuses on enhancing agroecosystem health through methods like crop rotation, intercropping, resistant cultivars, manure and compost addition, and reduced soil tillage. Despite its environmental benefits, organic agriculture often faces a yield gap compared to conventional methods, with an average decrease of 16%. This yield gap varies among crops and regions, highlighting the need for new crop cultivars and management techniques in organic agriculture (Richard *et al.*, 2021).

Conservation farming is another alternative system, aiming to improve soil health and limit erosion through principles like minimising soil disturbance, maintaining organic matter cover, and diversifying crop rotations. This approach is often associated with regenerative agriculture, focusing on sustainable practices. While low- and high-input agricultural systems share some practices, their independence varies. Farmers may choose to incorporate certain practices depending on their preferences. However, changes in fertilization or tillage practices can indirectly impact crop diseases. In terms of crop protection, these systems rely on Integrated Crop Management (ICM) or Integrated Pest Management (IPM) to minimise yield loss due to diseases. ICM principles involve preventive or suppression methods, monitoring pests and pathogens, decision support tools, non-chemical control methods, and targeted pesticide use with minimal side effects. These principles, emphasising resistance, avoidance, or elimination of pathogens, are crucial for both low- and high-input agricultural systems and are applicable in conventional agriculture as well.

Different Types of Innovative Management Practices in Field Crops

1. **Integrated Pest Management (IPM):** IPM is a holistic approach that combines various strategies to control pests and diseases in a sustainable manner.

   Components: Biological control, cultural practices, mechanical control, and the judicious use of chemical pesticides are integrated to minimise environmental impact.

   Benefits: Reduces reliance on chemical pesticides, promotes natural predators, and enhances long-term sustainability.

2. **Precision Agriculture:** Precision agriculture utilises technology, such as drones, sensors, and data analytics, to optimise the use of resources in farming practices.

   Application: Targeted delivery of pesticides and fertilisers based on real-time data, minimising wastage and maximising efficiency.

   Benefits: Improved resource management, reduced environmental impact, and increased crop yields.

3. **Biological Control and Biopesticides:** Biological Control: Introducing natural predators or parasites to control pest populations (Sahoo *et al.*, 2023b).

   Biopesticides: Using microorganisms, botanical extracts, or biochemicals to control pests and diseases.

   Benefits: Environmentally friendly, reduced chemical usage, and minimal harm to non-target organisms.

4. **Gene Editing and Resistant Varieties:** CRISPR technology and other gene-editing tools enable the development of crops resistant to specific pests and diseases.

   Benefits: Reduced reliance on chemical interventions, increased crop resilience, and improved food security.

5. **Crop Rotation and Diversification:** Rotating crops disrupts the life cycles of pests and diseases, reducing their prevalence. Diversification involves planting multiple crops in the same area to create a more resilient and balanced ecosystem (Sahoo *et al.*, 2023c).

   Benefits: Natural pest and disease control, improved soil health, and increased biodiversity.

6. **Remote Sensing for Early Detection:** Utilising satellite imagery and remote sensing technologies to detect early signs of pest infestations or diseases.

Benefits: Early intervention, precise targeting of affected areas, and prevention of widespread crop damage.

7. **Farm-level Digital Platforms:** Adoption of digital platforms that provide real-time information on pest and disease outbreaks.

   Benefits: Enables proactive decision-making, facilitates information sharing among farmers, and supports community-based pest control initiatives.

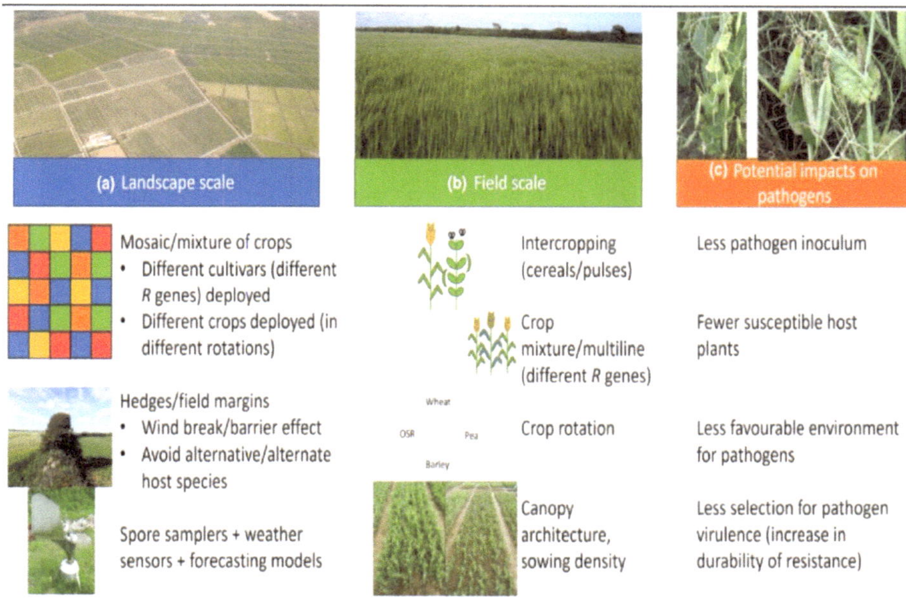

Fig. 1: Strategies for Integrated Crop Management aimed at reducing the severity of arable crop diseases can be implemented at either a landscape scale (a) or a crop field scale (b), considering their potential impacts on crop pathogen population dynamics (c).

Conclusion

Innovative management practices for diseases and pests in agriculture are crucial for the long-term sustainability of food production. By embracing technologies, ecological approaches, and data-driven strategies, farmers can not only protect their crops more effectively but also contribute to a healthier and more resilient agricultural ecosystem. The ongoing evolution of these practices holds promise for a future where agriculture is not only productive but also environmentally sustainable.

References

Barzman, M., Bàrberi, P., Birch, A. N. E., Boonekamp, P., Dachbrodt-Saaydeh, S., Graf, B., *et al.* (2015). Eight principles of integrated pest management. Agronomy for Sustainable Development, 35, 1199-1215.

Barzman, M., Bàrberi, P., Birch, A. N. E., Boonekamp, P., Dachbrodt-Saaydeh, S., Graf, B., *et al.* (2015). Eight principles of integrated pest management. Agronomy for Sustainable Development, 35, 1199-1215.

Cruces, L., Pena, E. D. L., & De Clercq, P. (2020). Seasonal phenology of the major insect pests of quinoa (Chenopodium quinoa Willd.) and their natural enemies in a traditional zone and two new production zones of Peru. Agriculture, 10(12), 644.

Göldel, B., Lemic, D., & Bažok, R. (2020). Alternatives to synthetic insecticides in the control of the colorado potato beetle (Leptinotarsa decemlineata) and their environmental benefits. Agriculture, 10(12), 611.

Kadoić Balaško, M., Mikac, K. M., Benítez, H. A., Bažok, R., & Lemic, D. (2021). Genetic and morphological approach for western corn rootworm resistance management. Agriculture, 11(7), 585.

Kuramoto, J., & Sagasti, F. (2002). Integrating local and global knowledge, technology and production systems: challenges for technical cooperation. Science, Technology and Society, 7(2), 215-247.

Lykogianni, M., Bempelou, E., Karamaouna, F., & Aliferis, K. A. (2021). Do pesticides promote or hinder sustainability in agriculture? The challenge of sustainable use of pesticides in modern agriculture. Science of the Total Environment, 795, 148625.

Poggi, S., Le Cointe, R., Lehmhus, J., Plantegenest, M., & Furlan, L. (2021). Alternative strategies for controlling wireworms in field crops: A review. Agriculture, 11(5), 436.

Richards, B., Qi, A., Fitt, B.D.L. (2021). Control of crop diseases through Integrated Crop Management to deliver climate-smart farming systems for low- and high-input crop production. Plant Pathology, 75(1), 187-206.

Sahoo, R., Sow, S., Ranjan, S., Samanta, D., & Sadhu, S. (2023a). Climate Resilient Strategies for Plant Disease Management. Agri-India TODAY, 3(2), 49-53.

Sahoo, R., Sow, S., Ranjan, S., Samanta, D., & Sadhu, S. (2023b). Effect of Climate Change in Plant Disease Development and Its Management Strategies. In: Recent Fronteers in Plant Science Research, Blue Duck Publications, Srinagar, J&K, pp. 32-41.

Sahoo, R., Sow, S., Ranjan, S., Samanta, D., & Sadhu, S. (2023c). Biorational Approach: An Eco-Friendly Approach to Control the Wilt Diseases of Crops. In: Pest Management Strategies in Pulses and Cereal Crops, Kripa-Drishti Publications, Pune, pp. 21-28.

Sow, S., & Ranjan, S. (2020). Integrated Weed Management in Rice. Agriculture & Food: E-Newsletter, 2(12), 70-73.

6

Interactions Between Agricultural and Environmental Policies

***Shirisha Junuthula*[1] *and Veenita Kumari*[2]**

[1]*Manage Fellow, National Institute of Agricultural Extension Management (MANAGE) Rajendranagar, Hyderabad 500030, India*

[2]*National Institute of Agricultural Extension Management, Rajendranagar Hyderabad 500030, India*

Abstract

The integration of agricultural and environmental policies stands as a critical imperative for sustainable development. This synthesis explores the complexities, challenges, and strategies for reconciling conflicting interests and achieving coherence between agricultural productivity and environmental conservation. Conflicting interests arise from divergent priorities in agriculture, emphasising productivity and economic growth, while environmental goals focus on conservation and sustainability. Economic, social, and political barriers further hinder policy coherence, necessitating holistic and adaptive approaches. Several strategies have been proposed to navigate these challenges. Integrated policy approaches, incentivising sustainable practices, and aligning economic incentives with environmental objectives emerge as promising avenues. Agroecological farming methods and ecosystem-based policies have gained traction, offering pathways to enhance biodiversity, soil health, and ecosystem resilience. However, their implementation requires farmer training, behavioural change, and overcoming quantification challenges for ecosystem benefits. Future policy developments should emphasise flexibility, education, and innovation. Adaptive governance, education programmes, and research investments are crucial for fostering awareness, driving behavioural changes, and advancing sustainable agriculture. This synthesis underscores the need for coherent policies that bridge the gap between agricultural productivity and environmental sustainability, steering towards a harmonious balance for long-term global well-being.

Keywords: *Agricultural policies, Biodiversity, Climate change, Environmental policies, Land use, Policy frameworks, Sustainability*

Introduction

The realm of agricultural policies and environmental policies stands as two critical pillars influencing the sustainability and equilibrium of our ecosystem. Agricultural policies primarily focus on regulations and measures instituted by governments or institutions to manage agricultural practices, production, distribution, and rural development. Conversely, environmental policies aim to safeguard and conserve natural resources, biodiversity, and ecosystems while mitigating the adverse impacts of human activities on the environment.

Overview of Agricultural and Environmental Policies

Agricultural policies often encompass subsidies, price controls, land use regulations, and trade agreements aimed at bolstering agricultural productivity, ensuring food security, and supporting rural economies (Fischer & Edmeades, 2010). Conversely, environmental policies encompass a broad spectrum of regulations, such as pollution control measures, wildlife protection, habitat conservation, and climate change mitigation strategies (Boyd & Banzhaf, 2007).

Importance of Understanding Their Interactions

Understanding the interactions between agricultural and environmental policies is crucial due to their interconnectedness and potential trade-offs. Agricultural activities significantly impact the environment through deforestation, water pollution from agricultural runoff, greenhouse gas emissions, and loss of biodiversity (Pretty *et al.*, 2008). Conversely, environmental conditions greatly influence agricultural productivity and sustainability, including soil quality, water availability, and climate patterns. Recognition of these interdependencies is vital for devising coherent policies that foster sustainable agricultural practices without compromising environmental integrity (Tilman *et al.*, 2011). The challenge lies in formulating policies that strike a balance between agricultural production goals and environmental conservation imperatives.

Agricultural and environmental policies play a crucial role in shaping sustainable development by addressing the complex relationship between agricultural practices and environmental preservation. These policies are formulated by governments and international bodies to ensure food security, preserve natural resources, mitigate climate change, and maintain biodiversity (Pretty, 2008).

Agricultural Policies

Agricultural policies encompass a wide range of measures aimed at supporting and regulating the agricultural sector. They include initiatives to enhance productivity, ensure fair market practices, provide subsidies, manage land use, and promote innovation and technology adoption in farming (Antle & Capalbo, 2001). These policies often influence land-use patterns and farming practices, impacting ecosystems and biodiversity.

Environmental Policies

Environmental policies focus on protecting and conserving the natural environment. They address issues such as air and water quality, soil conservation, biodiversity, and climate change mitigation. Measures to reduce pollution, promote renewable energy, and sustainably manage resources are central components of these policies (National Research Council, 2010).

Interactions between Agricultural and Environmental Policies

Land Use Impact: Agricultural policies influence land use by incentivising specific farming practices. For example, subsidies for monoculture farming can affect biodiversity and ecosystem health.

Water Resource Management: Agricultural practices heavily consume water resources. Policies regulating irrigation methods and water usage impact both agricultural productivity and water availability for other sectors and ecosystems.

Biodiversity Conservation: Agricultural activities can either promote or threaten biodiversity. Policies supporting sustainable farming practices and habitat preservation are crucial for maintaining diverse ecosystems.

Climate Change Mitigation: Both sets of policies contribute to climate change mitigation. Agricultural practices like agroforestry, incentivised through policies, can sequester carbon and reduce greenhouse gas emissions (Lipper *et al.*, 2014).

Importance of Understanding Interactions

Recognising the intricate connections between agricultural and environmental policies is critical. Unintended consequences, such as environmental degradation due to unsustainable agricultural practices or agricultural productivity constraints from stringent environmental regulations, highlight the need for a balanced approach (Pretty, 2008).

Understanding the interplay between agricultural and environmental policies is essential for policymakers to develop integrated strategies that optimise agricultural productivity while minimising adverse environmental impacts. Collaborative efforts and coherent policies are fundamental to achieving sustainable development and ensuring the well-being of present and future generations.

Impacts of Agricultural Policies on the Environment

Agricultural policies have significant impacts on the environment, particularly concerning land use and water management practices. These policies, including subsidies, incentives, and regulations, directly influence agricultural practices and subsequently affect the environment. Focusing on India, a country with a substantial agricultural sector, let's delve into the impacts of agricultural policies on the environment, particularly regarding land use and water management.

Effects of Subsidies, Incentives, and Regulations on Land Use

- Subsidies and Land Use: Subsidies in India often favour specific crops or farming practices, influencing land-use decisions. For instance, subsidies on fertiliser fertilisers might encourage excessive use, leading to soil degradation and water pollution (Kumar & Singh, 2018). Subsidies on particular crops might result in monoculture, reducing biodiversity and soil fertility.
- Incentives and Agroforestry: Government incentives for agroforestry or mixed-crop cultivation can positively impact land use. These practices promote biodiversity, prevent soil erosion, and enhance ecosystem services (Srivastava *et al.*, 2020). However, the extent of these incentives and their implementation significantly affects adoption rates.
- Regulations and Land Management: Land use regulations influence farming practices. For instance, regulations on land fragmentation can encourage consolidation, potentially leading to larger-scale mechanised farming, affecting local ecosystems and traditional farming practices.

Water Management and Conservation Practices

- Irrigation Subsidies: Subsidies on irrigation in India have led to unsustainable groundwater extraction and inefficient water use (Shah, 2009). These subsidies often do not encourage water conservation practices, contributing to water scarcity and depletion of aquifers.
- Water Conservation Policies: Policies promoting water conservation practices, such as rainwater harvesting or drip irrigation, are crucial. Some states in India have initiated programmes incentivising water-saving techniques, but their effectiveness varies (Bhattacharya *et al.*, 2020).
- Regulation on Pesticide and Fertiliser Use: Regulations on pesticide and fertiliser use are pivotal for water quality. Excessive use without proper regulation can lead to water pollution, affecting aquatic ecosystems and human health (Kumar & Singh, 2018).

The impacts of agricultural policies on the environment in India are significant. Subsidies, incentives, and regulations play pivotal roles in shaping land use, water management, and conservation practices (Anuragi *et al.*, 2023; Ranjan and Sow, 2020). While certain policies might promote sustainable practices, others can lead to environmental degradation if not carefully implemented or monitored. A balanced approach that encourages sustainable land use and water management practices is crucial for India's agricultural sector's long-term viability and environmental conservation (Alhammad *et al.*, 2023).

Influence Environmental Policies on Agriculture

Environmental policies have a significant influence on agriculture, especially in a country like India, where agriculture is a crucial sector supporting millions of livelihoods. Let's explore how environmental policies impact farming practices, biodiversity conservation, and ecosystem services in India (Sow *et al.*, 2021).

Impact of Environmental Regulations on Farming Practices

Pesticide and Chemical Regulation: Environmental regulations in India regarding pesticide and chemical usage aim to protect human health and the environment. Strict regulations on the use of harmful pesticides and chemicals influence farming practices. Farmers are encouraged or mandated to adopt integrated pest management (IPM) practices, organic farming, or reduced chemical usage (Sharma *et al.*, 2019).

Soil and Water Conservation Measures: Environmental policies promoting soil conservation and water management impact farming practices. Policies emphasising soil health, erosion control, and efficient water use encourage farmers to adopt conservation agriculture techniques, such as zero-tillage or contour farming (Jat *et al.*, 2019).

Climate Change Mitigation Policies: Policies targeting climate change mitigation often influence farming practices. In India, initiatives supporting climate-smart agriculture encourage the adoption of practices like agroforestry, crop diversification, and water-saving technologies to reduce greenhouse gas emissions and enhance resilience (Mishra *et al.*, 2020). Table 1 provides a comparative analysis of different policy approaches, outlining their key features, advantages and challenges in integrating agriculture and environmental conservation.

Table 1. Comparative Analysis of Policy Approaches

Policy Approach	Key Features	Advantages	Challenges
Agro-ecological	Promotes biodiversity	Enhances soil health and resilience	Requires farmer training and behavioural change
Ecosystem-based	Focuses on ecosystem	Enhances ecosystem services provision	Difficulty in quantifying ecosystem benefits
Integrated Approach	Holistic strategies	Considers both agricultural and environmental aspects	Requires inter-sectoral collaboration
Incentive Alignment	Aligns economic incentives	Encourages adoption of sustainable practices	Costly implementation and monitoring

Biodiversity Conservation and Ecosystem Services

Biodiversity Policies: India has various policies focused on biodiversity conservation. Protected area networks, biodiversity hotspots, and biosphere reserves are part of these initiatives. Policies promote sustainable agricultural practices within and around these areas, safeguarding biodiversity while supporting local agriculture (Gupta *et al.*, 2019).

Ecosystem Services Recognition: Environmental policies increasingly recognise the importance of ecosystem services provided by agricultural landscapes. Policies supporting ecosystem-based approaches aim to enhance pollination services, soil fertility, water purification, and carbon sequestration, encouraging sustainable farming practices (Sekar *et al.*, 2019). Environmental policies in India significantly influence agricultural practices, biodiversity conservation, and ecosystem services. Regulations and incentives play pivotal roles in steering farmers towards sustainable practices that benefit both agricultural productivity and environmental conservation. However, effective implementation, awareness, and support mechanisms are essential to ensure the success and widespread adoption of these policies among farmers.

Case Studies and Policy Frameworks

Case studies showcasing successful policies integrating agriculture and the environment and analyse international agreements such as the Paris Agreement and their implications on agriculture.

Examples of Successful Policies Integrating Agriculture and Environment

China's Grain-to-Green Programme (GTGP): Initiated in 1999, this programme aimed to mitigate soil erosion and land degradation by converting croplands on steep slopes to forests or grasslands. It successfully restored ecological balance, improved biodiversity, and enhanced ecosystem services while ensuring food security (Zhang *et al.*, 2018).

Brazil's Forest Code: The Brazilian Forest Code mandates that a certain percentage of private land must be maintained as forest cover. It balances agricultural expansion with forest conservation, contributing to biodiversity protection, carbon sequestration, and sustainable land use (Griscom *et al.*, 2020).

Sikkim's Organic Farming Initiative (India): Sikkim became India's first fully organic state by implementing policies to promote organic farming practices. This initiative prioritised sustainability, reduced chemical inputs, conserved biodiversity, and enhanced soil health (Bhaskar *et al.*, 2017).

European Union's Common Agricultural Policy (CAP): The CAP emphasises sustainable farming practices, promoting biodiversity conservation and addressing environmental challenges. Through mechanisms like agri-environmental schemes, the CAP provides financial incentives to farmers practicing environmentally friendly methods such as organic farming, crop diversification, and habitat preservation. It aims to balance agricultural production with environmental conservation and has contributed significantly to habitat preservation and biodiversity enhancement (European Commission, 2023).

Australia's Reef 2050 Plan: The Reef 2050 Plan focuses on the conservation and sustainable management of the Great Barrier Reef, recognising agriculture as a significant contributor to water quality issues affecting the reef. The plan includes measures to reduce agricultural runoff, improve soil health, and enhance land management practices to protect the reef's ecosystem. It involves partnerships with farmers, industry, and government agencies to implement sustainable agricultural practices while safeguarding the reef's health (Great Barrier Reef Marine Park Authority, 2021).

Kenya's Green Belt Movement: The Green Belt Movement, founded by Nobel Peace Prize laureate Wangari Maathai, combines afforestation initiatives with community development and women's empowerment. It encourages tree planting to combat deforestation, restore degraded landscapes, improve soil fertility, and mitigate climate change. This movement integrates environmental conservation with livelihood improvement, promoting sustainable agriculture and fostering community resilience (Green Belt Movement, 2023).

These examples illustrate how diverse policies across different regions emphasise the integration of agriculture and environmental conservation.

They showcase approaches that incentivise sustainable practices, address environmental challenges, and engage stakeholders to achieve both agricultural productivity and ecological sustainability.

Analysis of International Agreements and Agricultural Implications

Paris Agreement: The Paris Agreement focuses on mitigating climate change and adapting to its impacts. While not directly addressing agriculture, it acknowledges the sector's role in emissions and adaptation. Countries are encouraged to implement policies supporting climate-smart agriculture, sustainable intensification, and resilience-building in farming practices (FAO, 2018).

UN Sustainable Development Goals (SDGs): Several SDGs directly relate to agriculture and the environment, emphasising the need for sustainable food production, climate action, biodiversity conservation, and land restoration. The SDGs provide a framework for aligning agricultural policies with environmental and developmental objectives on a global scale (United Nations, 2023).

Successful policies integrating agriculture and the environment demonstrate the feasibility of balancing agricultural productivity with environmental conservation. International agreements like the Paris Agreement and the SDGs recognise agriculture's significance in environmental sustainability and encourage nations to align their policies to achieve these global objectives.

Convention on Biological Diversity (CBD): The CBD emphasises the conservation of biological diversity, the sustainable use of its components, and the fair and equitable sharing of benefits arising from genetic resources. It has significant implications for agriculture by encouraging the conservation and sustainable utilisation of crop genetic resources, promoting agro-biodiversity, and supporting traditional farming practices. The CBD's implementation involves strategies to maintain genetic diversity in agricultural systems, safeguard ecosystems, and enhance food security (CBD, 2023.).

World Trade Organisation's Agreement on Agriculture (AoA): The AoA aims to create fair competition in agricultural trade and ensure food security. It has implications for agricultural policies, subsidies, and market access. The agreement's impact on agriculture and the environment varies, with some policies leading to intensification and specialisation while others support sustainable practices. Balancing trade liberalisation with environmental sustainability remains a challenge, highlighting the need for policy coherence across trade, agriculture, and the environment (WTO, 2020).

United Nations Framework Convention on Climate Change (UNFCCC): The UNFCCC addresses climate change and its impacts, recognising agriculture's dual role as a contributor to emissions and a sector vulnerable to climate change. International climate agreements under the UNFCCC, such as the Paris Agreement, encourage countries to adopt climate-smart agricultural practices to mitigate greenhouse gas emissions, enhance resilience, and adapt to changing climatic conditions. These agreements promote sustainable land use, agroforestry, and agricultural practices that sequester carbon and improve soil health (UNFCCC, n.d.).

International agreements play significant roles in shaping agricultural policies, trade relations, and environmental sustainability. They offer frameworks and guidelines that influence countries' approaches to agriculture, encouraging practices that balance productivity, trade, and environmental conservation on a global scale.

Challenges in Policy Integration

Integrating agricultural and environmental policies poses several challenges due to conflicting interests, trade-offs, and barriers across economic, social, and political dimensions. These challenges hinder the achievement of coherent policies that effectively balance agricultural productivity with environmental sustainability.

Conflicting Interests and Trade-offs between Agricultural and Environmental Goals

Competing Priorities: Agricultural policies often prioritize increased productivity, food security, and economic growth. However, these goals might conflict with environmental conservation priorities such as biodiversity preservation, soil health, and water quality (Giller *et al.*, 2017).

Land Use Conflicts: Land serves multiple purposes—agricultural production, urban development, conservation, etc. Balancing these conflicting land use demands while minimising habitat loss, deforestation, and ecosystem degradation is challenging (Fischer *et al.*, 2015).

Resource Allocation Dilemmas: Resource allocation between agriculture and environmental conservation can be contentious. For instance, water allocation for irrigation versus maintaining environmental flow for ecosystems creates dilemmas (Thompson *et al.*, 2019).

Economic, Social, and Political Barriers to Policy Coherence

Economic Pressures: Economic considerations often drive policy decisions. Agricultural profitability might be prioritised over environmental concerns due to the immediate economic benefits, leading to unsustainable practices (Scherer *et al.*, 2018).

Social Dynamics: Societal expectations, livelihoods of farmers, and consumer demand for affordable food often conflict with sustainable agricultural practices. Resistance to change and lack of awareness among stakeholders can impede policy coherence (Lubell *et al.*, 2016).

Political Challenges: Conflicting political interests, lobbying by powerful agricultural or industrial groups, and short-term political cycles can hinder the adoption of long-term sustainable policies. This leads to policy fragmentation and a lack of continuity (Jordan *et al.*, 2018).

The challenges in policy integration between agriculture and the environment stem from the inherent conflicts between differing goals, resource allocation dilemmas, and socio-political dynamics. Achieving policy coherence requires addressing these conflicts, balancing trade-offs, fostering multi-stakeholder engagement, enhancing awareness, and aligning economic incentives with environmental objectives.

Strategies for Effective Policy Integration

Effectively integrating agricultural and environmental policies requires strategies that reconcile conflicting interests and promote synergy between agriculture and the environment. Here are some approaches and policy recommendations for achieving policy coherence:

Approaches to Reconcile Conflicting Interests

- Integrated Approach: Implement integrated policies that consider both agricultural and environmental objectives simultaneously. This involves fostering collaboration between relevant sectors, government agencies, stakeholders, and experts to develop holistic strategies (Scoones *et al.*, 2020).
- Adaptive Governance: Adopt adaptive governance approaches that allow for flexible policy adjustments based on evolving knowledge, changing conditions, and stakeholder feedback. This promotes adaptive management and continuous improvement (Dietz *et al.*, 2020).
- Incentive Alignment: Align economic incentives with environmental goals. Offer financial support, subsidies, or tax incentives to farmers who adopt sustainable practices, promoting the adoption of environmentally friendly methods (D'Odorico *et al.*, 2018).

Suggestions for Future Policy Developments and Implementations

Holistic Approaches: Future policies should adopt holistic, integrated approaches that bridge the gap between agricultural productivity and environmental sustainability. This requires collaboration between sectors, stakeholders, and government bodies.

Flexibility and Adaptation: Policies should be flexible and adaptable, capable of responding to evolving knowledge, changing conditions, and stakeholder needs. Adaptive governance models can enable continuous improvements in policy design and implementation.

Incentivising Sustainability: Aligning economic incentives with environmental objectives is crucial. Future policies should offer incentives and support mechanisms for farmers adopting sustainable practices, promoting their widespread adoption.

Education and Awareness: Invest in educational programmes targeting farmers, policymakers, and the public. Increasing awareness about the benefits of sustainable agriculture and environmental conservation can drive behavioural changes and garner broader support for integrated policies.

Research and Innovation: Encourage research and innovation in sustainable agriculture. Invest in technological advancements and best practices that enhance productivity while minimising environmental impact.

Integrating agricultural and environmental policies remains a complex yet imperative task for achieving sustainable development. Addressing conflicting interests, overcoming barriers, and adopting integrated strategies are pivotal for policy coherence. Future policy developments should emphasise holistic approaches, flexibility, incentivisation of sustainability, education, and innovation to ensure a harmonious balance between agricultural productivity and environmental conservation.

By adopting these recommendations, policymakers can pave the way for more effective and coherent policies that support both agricultural growth and environmental sustainability.

Conclusion

Integrating agricultural and environmental policies faces challenges due to conflicting goals. Agriculture emphasises productivity and economic growth, while environmental goals focus on conservation and sustainability, leading to trade-offs. Economic, social, and political factors create barriers to policy coherence. Economic pressures prioritize short-term gains over long-term environmental sustainability. Social dynamics and political interests often hinder the adoption of integrated policies. Thus, several strategies such as adopting an integrated approach, adaptive governance, and aligning economic incentives are crucial for reconciling conflicting interests and achieving policy coherence.

References

Alhammad, B. A., Roy, D. K., Ranjan, S., Padhan, S. R., Sow, S., Nath, D., Seleiman, M. F., & Gitari, H. (2023). Conservation Tillage and Weed Management Influencing Weed Dynamics, Crop Performance, Soil Properties, and Profitability in a Rice–Wheat–Greengram System in the Eastern Indo-Gangetic Plain. Agronomy, 13, 1953.

Anuragi, K., Bhavya, Sow, S., Ranjan, S., Dayal, P., & Nath, D. (2023). Conservation Agriculture as a Climate Smart Technology for Crop Production. In: Climate Smart Approaches towards Sustainable Crop Production. Vital Biotech Publication, Kota, India, pp. 77-88.

Antle, J. M., & Capalbo, S. M. (2001). Econometric-process models for integrated assessment of agricultural production systems. American Journal of Agricultural Economics, 83(2), 389-401.

Bhaskar, V., Gupta, S., & Hazra, K. K. (2017). Organic Farming in Sikkim: Revolution, Constraints, and Opportunities. International Journal of Agriculture, Environment and Biotechnology, 10(5), 549-558.

Bhattacharya, A., Chakraborty, D., & Harsha, K. H. (2020). Adoption of Micro Irrigation Systems in India: Policy Initiatives and Implications. Water, 12(9), 2453.

Boyd, J., & Banzhaf, S. (2007). "What are ecosystem services? The need for standardised environmental accounting units." Ecological Economics, 63(2-3), 616-626.

CBD. (2023.). Convention on Biological Diversity. Retrieved from https://www.cbd.int/

Dietz, T., Ostrom, E., & Stern, P. C. (2020). The struggle to govern the commons. Science, 369(6503), 1263-1264.

D'Odorico, P., Davis, K. F., Rosa, L., Carr, J. A., Chiarelli, D., Dell'Angelo, J. & Seekell, D. (2018). The global food-energy-water nexus. Reviews of Geophysics, 56(3), 456-531.

European Commission. (2023). Common Agricultural Policy. Retrieved from https://ec.europa.eu/info/food-farming-fisheries/key-policies/common-agricultural-policy_en

FAO. (2018). Climate change and food security: Risks and responses. Food and Agriculture Organisation of the United Nations.

Fischer, J., Abson, D. J., Butsic, V., Chappell, M. J., Ekroos, J., Hanspach, J. & Wehrden, H. V. (2015). Land sparing versus land sharing: moving forward. Conservation Letters, 8(5), 1-7.

Fischer, R.A., & Edmeades, G.O. (2010). "Breeding and cereal yield progress." Crop Science, 50(2), S85-S98.

Giller, K. E., Andersson, J. A., Corbeels, M., Kirkegaard, J., Mortensen, D., Erenstein, O & Stirling, C. (2017). Beyond conservation agriculture. Frontiers in Plant Science, 8, 277.

Great Barrier Reef Marine Park Authority. (2021). Reef 2050 Plan. Retrieved from https://www.gbrmpa.gov.au/our-work/reef-2050-plan

Green Belt Movement. (2023). Home. Retrieved from https://www.greenbeltmovement.org/

Griscom, B. W., Adams, J., Ellis, P. W., Houghton, R. A., Lomax, G., Miteva, D. A & Woodbury, P. (2020). Natural climate solutions. Proceedings of the National Academy of Sciences, 117(51), 32433-32442.

Gupta, N., Sharma, A., & Sharma, R. (2019). India's Biodiversity Conservation and Environmental Protection Laws: The Ecosystem and Global Perspectives. Biodiversity Conservation and Utilisation in a Diverse World, 25-40.

Jat, M. L., Sapkota, T. B., Khatri-Chhetri, A., & Jat, R. K. (2019). Policies, strategies, and impact of conservation agriculture in India. Renewable Agriculture and Food Systems, 34(2), 111-122.

Jordan, A., Huitema, D., van Asselt, H., Rayner, T. J., Schoenefeld, J. J., Tosun, J., & Forster, J. (2018). The governance of climate change adaptation: coherence across scales. Policy Sciences, 51(3), 389-405.

Kumar, P., & Singh, R. D. (2018). The role of government subsidies in Indian agriculture. Global Journal for Research Analysis, 7(10), 1-3.

Lipper, L., Thornton, P., Campbell, B. M., Baedeker, T., Braimoh, A., Bwalya, M., & Mann, W. (2014). Climate-smart agriculture for food security. Nature Climate Change, 4(12), 1068-1072.

Lubell, M., Niles, M. T., Hoffman, M., Herrick, J. E., Robertson, G. P., & Jackson, L. E. (2016). Sustaining agriculture in the American Midwest: an assessment of alternative practices. Agriculture, Ecosystems & Environment, 218, 163-170.

Mishra, A., Rai, P. K., & Gupta, R. D. (2020). Climate-Smart Agriculture (CSA) in India: Policy Perspective. Climate Change and Agriculture: Adaptation Strategies and Mitigation Options, 259-276.

National Research Council. (2010). Toward sustainable agricultural systems in the 21st century. National Academies Press.

Pretty, J. (2008). Agricultural sustainability: concepts, principles and evidence. Philosophical Transactions of the Royal Society B: Biological Sciences, 363(1491), 447-465.

Pretty, J., *et al.* (2008). "Resource-conserving agriculture increases yields in developing countries." Environmental Science & Technology, 42(12), 3999-4004.

Ranjan, S., and Sow, S. (2020). Zero Tillage as a Resource Conserving Technology in Rice -Wheat Cropping System. Vigyan Varta, 1(4), 35-37.

Scherer, L., De Oliveira Junior, J. F., & Fracasso, J. L. (2018). Sustainable agriculture and economic development in the rural environment. Revista de Economia e Sociologia Rural, 56(2), 331-350.

Scoones, I., Stirling, A., Abrol, D., Atela, J., Charli-Joseph, L., Eakin, H., *et al.* (2020). Transformations to sustainability: combining structural, systemic and enabling approaches. Current Opinion in Environmental Sustainability, 42, 65-75.

Sekar, I., Sudharsan, D., & Ramachandran, A. (2019). Ecosystem services mapping for prioritization of agricultural landscapes in India. Ecosystem Services, 40, 101041.

Shah, T. (2009). Taming the Anarchy: Groundwater Governance in South Asia. Resources for the Future. Washington, DC.

Sharma, A., Pandey, K. K., & Kumar, S. (2019). Environmental Legislation and Policies for Sustainable Agriculture in India. Journal of Pharmacognosy and Phytochemistry, 8(6), 3197-3200.

Sow, S., Ranjan, S., and Sarita. (2021). Agricultural Diversification: An Environment Friendly Approach for Sustainable Production. In: rop Diversification and Soil Health Management for Sustainable Development, Gene-Tech Book, pp. 46-55.

Srivastava, R. K., Rathore, A., & Singh, S. K. (2020). Agroforestry in India: A Perspective on Policies, Research, and Opportunities. Agroforestry Systems, 94(4), 1389-1399.

Thompson, L. M., Barclay, K., & Skousen, J. G. (2019). Agricultural water use and the water footprint in the United States. International Journal of Water Resources Development, 35(4), 543-555.

Tilman, D. *et al.* (2011). "Global food demand and the sustainable intensification of agriculture." Proceedings of the National Academy of Sciences, 108(50), 20260-20264.

UNFCCC. (2023). United Nations Framework Convention on Climate Change. Retrieved from https://unfccc.int/

United Nations. (2023). Sustainable Development Goals. Retrieved from https://sdgs.un.org/goals

WTO. (2020). Agreement on Agriculture. Retrieved from https://www.wto.org/

7

Indigenous Agricultural Practices for Sustainable Crop Production

Parveen Kumar[1] and Raj Kumar Gupta[2]

[1]SMS (Ag. Extension), KVK Ramban, SKUAST-Jammu, Jammu & Kashmir, India

[2]KVK Ramban, SKUAST-Jammu, Jammu & Kashmir, India

Abstract

Indigenous Knowledge Systems represent an accumulated reservoir of beliefs and wisdom passed down through generations by cultural transmission. It encompasses the understanding of the relationships between living beings, including humans, and their environment. This indigenous technical knowledge holds particular significance in the agricultural sector, contributing to its sustainability and profitability. Within agriculture, this knowledge is encapsulated in Indigenous agricultural knowledge (IAK), forming an unwritten body of knowledge with no systematic record of its principles, operations, or applications. The erosion of indigenous wisdom, especially in agriculture, is becoming increasingly evident, with the younger generation often lacking awareness of these practices. It is imperative to raise awareness among the youth about the economic and environmental benefits associated with indigenous agricultural practices, ensuring the preservation and continuation of this invaluable knowledge.

Keywords: *Indigenous agricultural knowledge, Cultural practices, Sustainable agriculture*

Introduction

Throughout their existence on this planet, human beings have adapted to the diverse environments they inhabit. Depending on the topography and prevailing natural conditions, individuals have successfully navigated the challenges posed by nature and their surroundings. As life has progressed, humans have gained insights into nature, its resources, and the prudent utilisation thereof. This understanding encompasses various natural phenomena, the associated challenges, and the methods to mitigate them. Over time, humans have evolved into repositories of such knowledge, collectively known as 'indigenous knowledge' or 'traditional wisdom,' passed down from one generation to the next. Each specific region possesses its unique set of indigenous knowledge reflecting the harmonious coexistence of humans with their environment.

Indigenous agricultural knowledge (IAK) stands as a crucial force in the realm of agricultural development. It represents a systematic and collective body of knowledge, practices, and beliefs that evolve through adaptation processes and are transmitted across generations through cultural channels, elucidating the intricate relationship between living beings and their environment. With this collective wisdom, farmers have adeptly cultivated food crops in diverse environmental conditions and seasonal variations, relying on indigenous techniques in the absence of external inputs, resources, and scientific knowledge. This includes refraining from the use of synthetic agrochemicals and instead embracing practices like crop rotations and soil fertility restoration through closed nutrient cycles (Melash *et al*., 2023).

Agronomic practices rooted in indigenous knowledge, such as inter-cropping, terracing, mixed cropping with legumes to enhance reliance on biological fertility, and the adoption of agroforestry, are pivotal aspects of traditional farming methods. These practices have the potential to significantly reduce environmental pollution by minimising the use of modern agricultural inputs like chemical fertilisers, insecticides, and herbicides, thereby mitigating carbon, nitrogen, and water footprints. This shift towards indigenous technologies and knowledge in modern agriculture, facilitated by intensified interactions within local networks and organisational arrangements, holds substantial importance in fostering sustainable agricultural practices.

Indigenous Knowledge

The word 'indigenous' is used to or refer to the people who originally lived in a place, rather than people who moved there from somewhere else. Indigenous Knowledge (IK) can be characterised as the collective knowledge held by people indigenous to a specific geographical region. Indigenous Knowledge is 'unique, traditional and local knowledge existing within and developed around specific conditions of women and men indigenous to particular geographic area (Grenier, 1998)'. The indigenous knowledge system is described as an accumulated repository of knowledge and beliefs transmitted through cultural means across generations. This knowledge pertains to the interactions among living beings, including humans, and their environment (Dissanayake *et al*, 2012).

Native knowledge, traditional knowledge, cultural knowledge, and civilisation knowledge are interchangeable terms. It is unique to a given culture, society or country. Indigenous Knowledge is 'unique, traditional and local knowledge existing within and developed around specific conditions of women and men indigenous to a particular geographic area. The indigenous knowledge system is described as an accumulated repository of knowledge and beliefs that are transmitted through cultural means across generations. It encompasses understanding the relationships among living beings, including humans, with one another and their environment. This indigenous technical knowledge holds greater relevance in the agricultural sector to make it more sustainable and profitable. In agriculture, all such knowledge comes under Indigenous Agricultural Practices (IAP). IAP is an unwritten body of knowledge. There is also no systematic record of what they are, what they do and how they do what they do, how they can be changed, their operations, their boundaries and their applications.

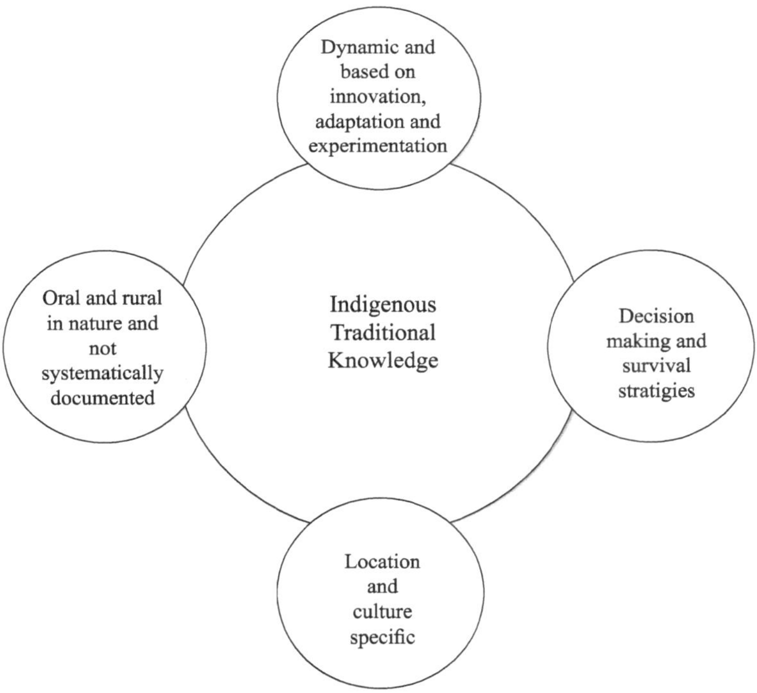

Fig. 1: Basic features of indigenous traditional knowledge

Indigenous Knowledge and Tribal Communities

The tribal communities residing in distant, remote, and hard-to-reach areas, contending with unpredictable weather conditions, serve as repositories of Indigenous Knowledge. About 8.6 percent of the Indian population belongs to a category listed as 'Scheduled Tribes' enumerated in the Schedule to Article

342 of the Constitution of India (Dutta Prabhash. 2022). Tribal people have been seen to be strongly associated with the forests, hills and remote areas, practicing a unique lifestyle and having a unique set of cultural and religious beliefs. Almost 90% of the tribal population of the country lives in rural areas. There are 90 districts or 809 blocks with more than 50% tribal population and they account for nearly 45% of the Scheduled Tribe (ST) population in the country. To clarify, around 55% of the tribal population lives outside of these 809 tribal-majority blocks. In 2007, an independent study disclosed that the tribal homelands in Central India, spanning approximately 100 districts nationwide, are home to an estimated 55 million tribal individuals, accounting for over 70% of India's tribal population. Despite the lush vegetation and ample rainfall in this region, it has one of the highest concentrations of rural poverty in the world. Over centuries, these communities have developed diverse indigenous agricultural practices for growing different crops, processing and storing produce, and conserving natural resources.

Characteristics of ITK

The major characteristics of ITK include

a. ITK is dynamic in nature
b. ITK is recorded orally and transferred from generation to generation
c. ITK is learned through observation and hands-on experience
d. ITK is mainly qualitative in nature ad needs a holistic approach
e. ITK has the potential to provide valuable insights into local resources, opportunities, challenges and possibilities

Importance of ITK

The development of indigenous agricultural practices, including the management of the natural environment has been a matter of survival for the people who have developed these systems. The traditional environmental knowledge of the people of India plays a significant role in formulating location-specific strategies for biodiversity protection and is valuable for creating mitigation plans to address climate change and ensure sustainability. The insights gained from close interactions with natural and physical environments, coupled with cultural adaptation, are increasingly acknowledged as more eco-friendly and sustainable. Additionally, this local knowledge and associated practices have been documented to assist in drought prediction and the management of extreme weather conditions (Sow and Ranjan, 2021).

ITK for Sustainable Agriculture

Contemporary policies and programs are centered around sustainability, with organic agricultural methods forming the core of sustainable farming practices. These methods predominantly avoid the use of toxic chemicals, which have inflicted irreversible damage on the ecosystem. The current philosophy advocates employing chemical control methods, such as insecticides and pesticides, on plants only as a last resort, after exhausting other cultural and preventive measures. If the use of an insecticide or pesticide becomes necessary, the preference is for the least toxic option. Several pesticides can be concocted at home using commonly available ingredients, presenting advantages such as accessibility, cost-effectiveness, and low toxicity to humans and other mammals. Examples like garlic solution, neem oil, pepper, soap solution, and tobacco solution showcase how traditional wisdom can be harnessed to eliminate chemicals from agriculture and protect the environment.

- Organic Agriculture: The government's promotion of Natural Farming is actually a combination of indigenous knowledge. The four components of Natural Farming, namely Beejamrit, Jeevamrit, Mulching, and Wapasha, are the result of indigenous wisdom accumulated through experience and practice. Various plant protection preparations like Neemastra, Brahmastra, Agniastra, and Dashparni extract used in natural farming also stem from local wisdom. Farmers can determine the optimal time for seed sowing by gauging soil moisture through a simple test, creating a soil ball from their fields.
- Multiple Cropping or Mixed Cropping: This cropping system involves the simultaneous cultivation of two or more crops in the same piece of land. The primary advantage of this approach lies in maintaining optimal soil conditions and controlling diseases and pests, among other benefits. The coexistence of crops provides a balance of light and shade, and the system allows for the cultivation of crops with different maturity periods. In indigenous practices, seeds are randomly sown by hand in the field; however, in mixed cropping, the seeds of all crops are sown simultaneously in the same field. Inter-planting enables cropping systems to recycle their own stored nutrients, leading to higher productivity per unit area compared to mono-cropping systems.
- Crop Rotation: This cropping system involves cultivating crops during the fallow period. Its key characteristic is the ability to cultivate a well-chosen combination of crops in a specific sequence over several years without depleting soil fertility. The primary goal of this system is to enhance and preserve soil fertility, prevent pest and soil-borne disease

infestations, control soil erosion, and more. Crop selection for rotation is based on fostering complementary or supplementary relationships among them rather than fostering competition (Ranjan and Sow, 2021).

- Fallowing: In numerous locations, fields are intentionally left fallow for a period of 2–5 years to allow the soil to naturally replenish its fertility. During the fallow period, domestic animals are allowed to graze on the fields, and their droppings contribute to enhancing soil fertility. Ecological succession plays a significant role in this process during the fallow period.

ITK for Storage of Perishables Commodities

Indigenous agricultural practices play a significant role, especially in regions where the climate poses limitations. These practices are particularly prevalent in areas where farmers must store produce for winter months due to freezing temperatures that inhibit outdoor cultivation. In the Union Territory of Ladakh, structures like Cellars/Tsothbang, Sadong, are employed for vegetable storage. The Cellar, an underground structure, is constructed by excavating pits with dimensions ranging from 3 to 5 meters in length, breadth, and depth, tailored to the farmer's storage needs for winter vegetables.

In *Gurez*, people have adopted a similar practice of digging underground pits *known* as 'Khow' to store potatoes and other vegetables (Kumar and Bhat, 2023). These methods are well-established, reliable, and operate without the need for electricity or external power sources, Utilising locally available resources. Furthermore, these traditional techniques incur minimal costs and require no additional energy. Studies indicate that the shelf life of vegetables stored in these structures is comparable to modern, energy-intensive storage methods. The stored vegetables serve as a crucial source of nutritious food during the snow-covered period in the region. Another important storage technique for onions in the union territory of Ladakh is the '*Charches*'. It involves hanging of onions from the ceiling of roofs. Onions are harvested as soon as the bulb gets mature but the leaves are still green and tied in bundles. Onions are made into bundles with each of them weighing 4-5 kilograms.

This is a parallel arrangement where it is ensured that the two bunches do not touch each other and there is no physical damage to the onions. The onions are taken out from these bunches from one end so that the whole lot does not get disturbed. A similar technique has also been reported from Odisha for the storage of onions.

ITK for Homemade Pesticides

Insect pests cause serious damage to crops and livestock. Indigenous Traditonal Knowledge (ITK) in the form of homemade pesticides offers a sustainable approach to protecting crops and the environment. Various homemade pesticides include Garlic Spray that works in several ways. Its odour is a repellent to some insects; concentrated garlic is fatal to soft-bodied insects; and it has antibacterial properties that prevent diseases. Individually or in combination, garlic, oil, and soap can be utilised in plant protection sprays. The soap solutions also provide fungicidal protection from diseases like black spot, canker, powdery mildew and rust, while oil sprays are used to prevent powdery mildew infection of roses and fruit trees. Capsaicin, the compound responsible for the heat in hot peppers, also acts as a deterrent for various garden pests. A hot pepper spray can be made by chopping several hot peppers of any variety, the hotter the better and steeping them in 2 cups of hot water. Once the liquid cools, strain it into a jar with a squirt or two of liquid soap to create a hot pepper concentrate. Mix 2 teaspoons of the concentrate into 2 cups of water for a spray. For treating plants infested with spider mites, mix two tablespoons of salt (crystal salt) into one gallon of warm water and spray on infected areas. Mineral oils also work as organic pesticides against insects/pests. These oils dehydrate insects and their eggs. They can be made by mixing 10 to 30 ml of high-grade organic mineral oil with one liter of water. The mixture should be stirred well and then added to spray bottles. Although Tobacco is hazardous to human beings, there was a time when tobacco spray was commonly used as a pesticide for killing pests, caterpillars, and aphids. To make tobacco spray, mix 1 cup of organic tobacco (preferably a brand that is organic and all-natural) into 1 gallon of water. Allow the mixture to sit overnight. After 24 hours, the mix should have a light brown color. If it is very dark, add more water. This mix can be used on most plants, except those in the Solanaceae family (tomatoes, peppers, eggplants, etc.). Similarly, red pepper powder can also be used to create a homemade pesticide by mixing 1 tablespoon of red pepper powder, 6 drops of dish soap and 1 gallon of water thoroughly. Pour the red pepper mixture in a garden sprayer and thoroughly cover the vegetables with the spray. If needed, reapply the spray once a week to keep garden pests such as leafhoppers, spittlebugs, beetles and loopers off the plants.

How Indigenous Knowledge of People Are Helping World Achieve Zero Hunger?

The first target of Sustainable Development Goal 2 (SDG 2.1) aims to achieve zero hunger (UN). It is one of the 17 Sustainable Development Goals established by the United Nations in 2015. It states that by 2030, the goal is to end hunger and ensure access for all , particularly the poor and those in vulnerable situations including infants to safe nutritious and sufficient food all year round. The indigenous agricultural practices are resilient to climate change, as they conserve and restore habitats, while also expanding and diversifying diets with indigenous foods. Indigenous territories hold 80% of the world's biodiversity. The lifestyles of indigenous peoples are locally adopted and respectful of natural resources.

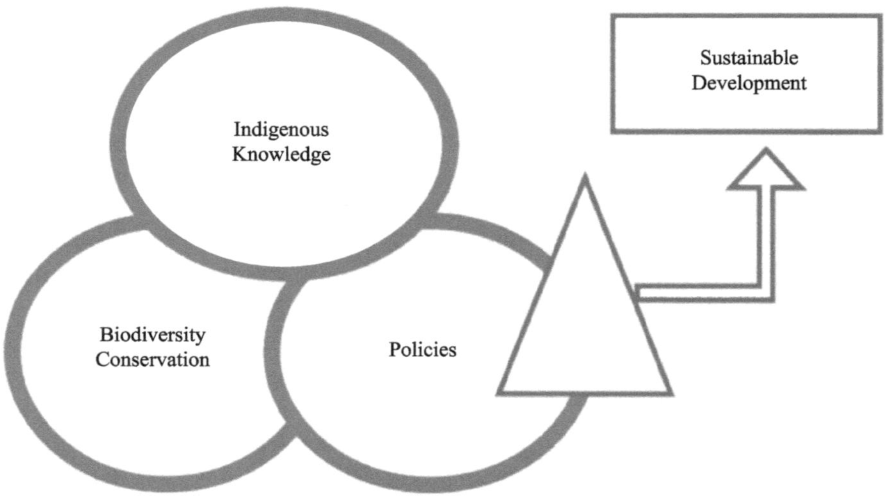

Fig. 2: Correlation between indigenous knowledge, biodiversity conservation, and different policies for sustainable development

Conclusion

Indigenous knowledge and biodiversity share a direct correlation, where the impact of one is observed in the other; the loss of either could lead to the disappearance of the other. The decline in biodiversity is a consequence of the erosion of indigenous knowledge, posing a threat to the survival of indigenous people and risking the loss of their knowledge, culture, language, etc. This rich knowledge is facing extinction due to factors like globalisation, poverty as a form of capability deprivation, exclusion from education and other facilities, and current environmental threats.

Various agricultural practices employed by native populations, whether directly or indirectly, contribute to the improvement of biodiversity. Preserving this unique indigenous knowledge is crucial, as many scientific discoveries have roots in it; its loss would directly impact the livelihood of civilisations.

Governments and non-governmental organizations have designed policies for biodiversity conservation. Indigenous agriculture emerges as a significant factor in maintaining biodiversity by managing soil, gene pools, and conserving native varieties. Therefore, robust policies are essential to conserve both indigenous knowledge and biodiversity. Shifting people's mindset towards spiritual education that emphasises important ethics becomes crucial in this context. Utilising native or indigenous agricultural systems and technologies can enhance the efficiency of indigenous varieties and biodiversity, ultimately contributing to the preservation of both indigenous knowledge and biodiversity for sustainable development.

References

Dissanayake, L., Kumaran, S.A.N.T.V. (2012). Indigenous Knowledge Systems. Kumaran Book House, pp.12-17.

Grenier, L. (1998). Working with Indigenous Knowledge, International Development Research Centre, Canada.

Kumar, P., and Bhat, B.A. (2023) Pits of Life: Khow of Gurez. The North Lines, Jammu and Kashmir. (https://thenorthlines.com/pits-of-life-khow-of-gurez/)

Melash, A. A., Bogale, A. A., Migbaru, A. T., Chakilu, G. G., Percze, A., Ábrahám, É. B., and Mengistu, D. K. (2023). Indigenous agricultural knowledge: A neglected human based resource for sustainable crop protection and production. Heliyon, 9(1).

Ranjan, S., and Sow, S. (2021). A way towards sustainable agriculture through zero budget natural farming. Food & Scientific Reports 2(3), 30-32.

Sow, S., and Ranjan, S. (2021). Indigenous Technical Knowledge (ITK) for Sustainable Agriculture in India. Agriculture & Food E-Newsletter 3(1), 31-35.

8

Legume Cultivation in India Challenges and Future Prospects

S. Ashokh Aravind[1], A. Surendhar[2], K. Arun Kumar[3] A. Ammaiyappan[1] and K. Nagarajan[1]

[1]*Department of Agronomy, Tamil Nadu Agricultural University, Coimbatore Tamil Nadu, India*

[2]*Centre for Plant Breeding and Genetics, Tamil Nadu Agricultural University Coimbatore, Tamil Nadu, India*

[3]*Department of Crop Physiology, Tamil Nadu Agricultural University Coimbatore Tamil Nadu, India*

Abstract

Legume cultivation in India is a complex yet promising journey marked by challenges and successes. Legumes, crucial for sustainable agriculture, belong to the Fabaceae family and uniquely harness nitrogen, impacting soil fertility, food security and the economy. With over 70 species cultivated, including chickpea, pigeonpea, lentil, mung bean, urd bean, soybean and groundnut, legumes offer economic and nutritional advantages, contributing to protein, fiber and essential nutrient needs. India's Silent Pulse Revolution has led to self-sufficiency in pulse production through indigenous and sustainable practices. Legume cultivation provides ecological benefits like nitrogen fixation, soil enhancement, weed control and biodiversity conservation. Agronomic practices tailored to India include variety selection, crop rotation, soil preparation, sowing and irrigation. Despite successes, challenges include abiotic and biotic stresses, postharvest losses, storage issues and market fluctuations. Strategies for improvement involve developing high-yielding varieties, promoting sustainable practices, educating farmers and supporting market development. Looking forward, legumes are positioned to drive sustainability, nutritional fortification, economic prosperity and environmental stewardship in India. In gastronomy, legumes enrich Indian cuisine with their versatility and nutritional value. The future holds a central role for legumes in sustainable agriculture, contributing to a harmonious blend of tradition and innovation.

Keywords: *Agricultural challenges, Legume cultivation, Pulse revolution, Sustainable agriculture*

Introduction

Embarking on the journey of legume cultivation in India reveals a nuanced landscape marked by both challenges and promising horizons. The agricultural scenario is akin to orchestrating a symphony, where legumes play a vital role, contributing melodies of resilience and significance. Unveiling this leguminous symphony in India is an exploration into the intricate interplay of obstacles and potential triumphs within the agricultural sector. As we delve into an overview of legumes, commonly known as pulse crops, their significance in Indian agriculture becomes apparent. Belonging to the Fabaceae family, these plants possess a distinctive feature - the ability to harness nitrogen from the atmosphere. This unique characteristic renders them pivotal components in sustainable agricultural systems, impacting not only food and nutritional security but also influencing the broader scope of soil fertility enhancement and economic prosperity. The journey through legume cultivation in India unfolds as a harmonious narrative, weaving together the challenges faced and the promising prospects that lie ahead in this agricultural symphony (Sow and Ranjan, 2021).

Diversity of Cultivated Legumes in India

India stands as a testament to the rich tapestry of legume diversity, cultivating over 70 distinct species across its expansive landscapes. Prominent among these legumes are:

- **Chickpea** *(Cicer arietinum):* Renowned as 'chana' in India, chickpea reigns as the most widely consumed pulse crop. Recognised for its rich protein, fiber and essential mineral content, it assumes a crucial role in Indian diets.
- **Pigeonpea** *(Cajanus cajan):* Commonly known as 'tuvar' or 'toor', pigeonpea stands as a versatile legume, finding utility as a pulse, vegetable, or fodder, showcasing its adaptability in various applications (Sarkar *et al.*, 2020).
- **Lentil** *(Lens culinaris):* Referred to as 'dal', in India, lentil holds a prominent place as a staple food, celebrated for its high protein and fiber content, contributing to dietary richness (Sharma and Shukla, 2014).
- **Mung Bean** *(Vigna radiata):* Recognised as 'moong dal' the mung bean exemplifies versatility, serving as a whole grain, sprout, or flour. It stands out as a valuable source of protein, fiber and vitamins.
- **Urd Bean** (Vigna mungo): Popularly known as 'black gram' or 'urad dal', the urd bean graces various culinary delights, presenting itself as a reliable source of protein, fiber and essential vitamins.

- **Soybean** *(Glycine max):* Beyond its status as a major oilseed crop, soybean emerges as a valuable protein source, finding extensive use in the production of edible oil, tofu and various soy-based products.
- **Groundnut** *(Arachis hypogaea):* Commonly referred to as peanuts, groundnut, a vital oilseed crop, offers a rich source of protein, fiber and healthy fats.
- **Fodder Legumes**: Including berseem, lucerne and stylo, fodder legumes play a crucial role in supporting livestock production. Not only do they provide nutritious forage, but they also contribute to soil fertility enhancement.

Economic and Nutritional Advantages of Legumes

The role of legumes extends beyond the agricultural field, significantly impacting the Indian economy and nutritional landscape. Acting as a livelihood source for millions of farmers, contributing to export revenues and fortifying food security, legumes offer a myriad of benefits:

- High Protein Content: Legumes emerge as an outstanding source of plant-based protein, essential for tissue development and repair.
- Dietary Fiber Riches: Abundant in dietary fiber, legumes promote digestive health, regulate blood sugar levels and contribute to a feeling of fullness.
- Essential Vitamins and Minerals: Legumes provide a spectrum of essential vitamins and minerals, including iron, zinc, calcium and folate, vital for diverse bodily functions.
- Health Rewards: Consumption of legumes correlates with a reduced risk of chronic ailments, including heart disease, diabetes and certain types of cancer.

As we embark on this exploration, the profound impact of legumes on both the agricultural and nutritional landscapes in India becomes increasingly evident, laying the groundwork for a comprehensive understanding of their cultivation challenges and future prospects.

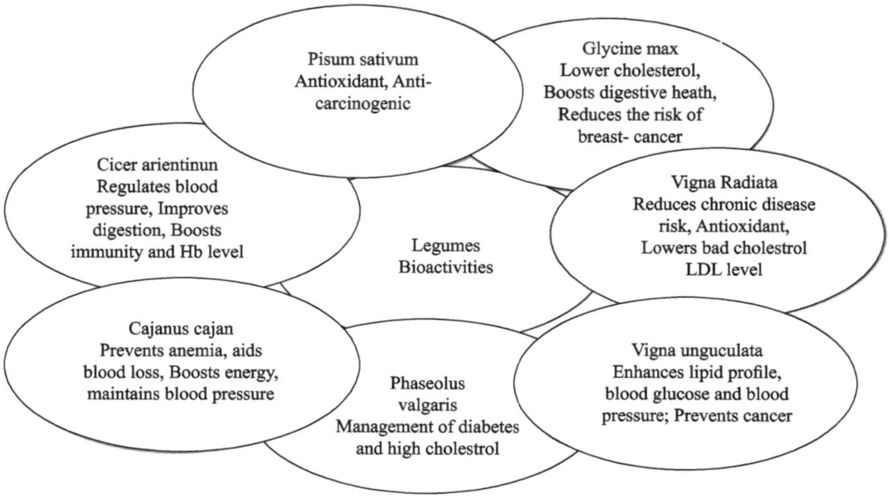

Fig. 1: Legumes: The powerhouses of nutrition (Singh *et al.*, 2022)

A. India's Silent Pulse Revolution: A Stride Towards Self-Sufficiency

Globally acclaimed for its significant share in the acreage and production of pulse crops, India stands as the largest consumer and processor of pulses worldwide. With over a dozen pulse crops cultivated, six major varieties take the forefront in the country. The recent years have witnessed a subtle yet impactful 'Pulses Revolution' in India. Notably, during the 2017–18 period, pulse production reached 25.23 million tonnes (mt), showcasing a substantial increase of 5.98 mt compared to 2013-14 (19.25 mt) and 2.10 mt over 2016–17 (23.13 mt). Over the last three consecutive Five-Year Plan periods, there has been a positive trajectory in the area, production and productivity of pulse crops. This upward trend indicates not only a significant increase in indigenous production but also its sustainability, positioning India on the path to self-sufficiency in pulses. In contrast to the renowned "Green Revolution" of 1968 that primarily focused on wheat and rice, the Pulses Revolution has silently unfolded, attributed to the development of science-led technologies and policy initiatives. Unlike its predecessor, the Pulses Revolution is distinctive for being led by indigenous technologies originating from research-based institutions. This realised revolution spans the nation, promoting environmentally friendly and sustainable practices, ultimately contributing to the country's food and nutritional security (Chaturvedi and Sandhu, 2020).

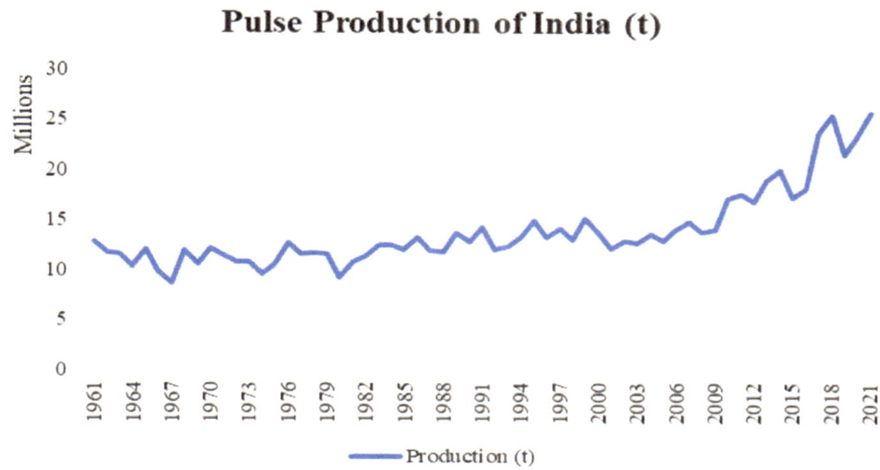

Fig. 2: Trend of pulse production in India from 1961 to 2021

Data source: UN Food and Agriculture OrganisationOrganisation (FAO). Note: Pulses are the edible seeds of plants in the legume family. The FAO recognizes 11 types of pulses: dry beans, dry broad beans, dry peas, chickpeas, cow peas, pigeon peas, lentils, Bambara beans, vetches, lupins and pulses (not elsewhere specified).

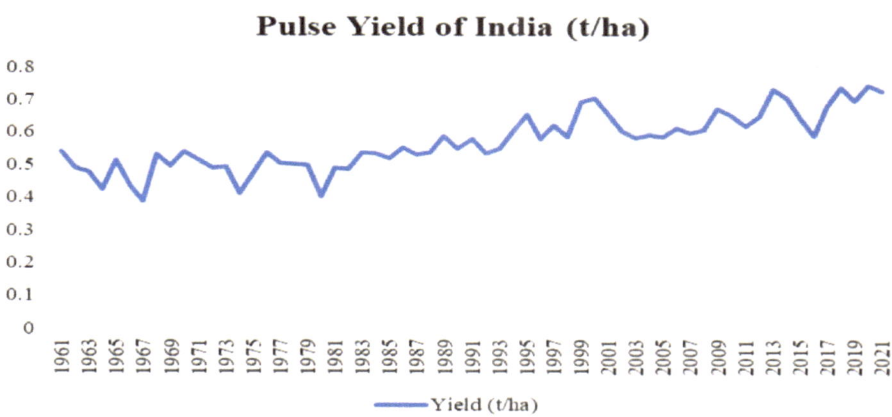

Fig. 3: Trend of pulse yield in India from 1961 to 2

Data source: UN Food and Agriculture OrganisationOrganisation (FAO). Note: Pulses are the edible seeds of plants in the legume family. The FAO recognizes 11 types of pulses: dry beans, dry broad beans, dry peas, chickpeas, cow peas, pigeon peas, lentils, Bambara beans, vetches, lupins and pulses nes (not elsewhere specified).

Ecological Benefits of Legume Cultivation in India

Continuing our exploration, we now delve into the ecological advantages bestowed by the cultivation of legumes in India. These benefits extend beyond the confines of agricultural practices, intertwining with the very fabric of environmental sustainability (Das and Ghosh, 2012).

A. Nitrogen Fixation and Soil Fertility Improvement

Arguably one of the most profound ecological contributions of legume cultivation lies in their unique ability to engage in nitrogen fixation. Within the intricate dance between legumes and their environment, symbiotic bacteria nestled in the root nodules of these plants perform a transformative feat - converting atmospheric nitrogen into a plant-friendly form. This alchemical process, known as nitrogen fixation, emerges as a game-changer, elevating soil fertility to new heights and concurrently diminishing the reliance on synthetic nitrogen fertilisers. As legumes gracefully contribute to this cycle, the agricultural landscape benefits from enriched soils, fostering a more sustainable and eco-friendly cultivation paradigm.

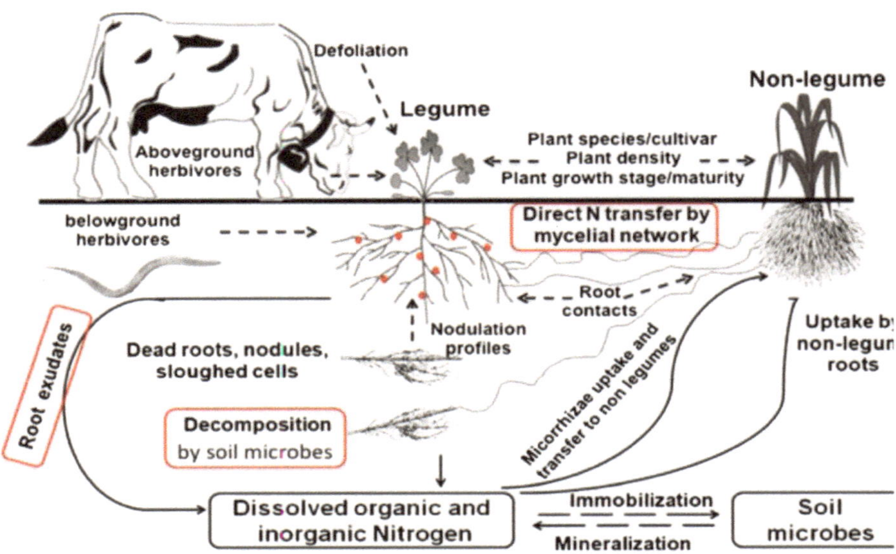

Fig .4: Exploring Below-Ground Nitrogen Transfers in Legumes to Non-Legumes: Unraveling Mechanisms and Influencing Factors (Thilakarathna *et al.*, 2016)

B. Weed Control and Crop Rotation

Beyond their nutritional contributions, legumes wield a formidable influence in the realm of weed management. Possessing allelopathic properties, these leguminous crops release natural chemicals that act as suppressors, inhibiting the growth of certain weeds. This intrinsic weed control characteristic not only reduces the dependence on herbicides but also promotes a more ecologically balanced and resilient agroecosystem. Furthermore, when strategically integrated into crop rotation practices, legumes emerge as allies in breaking disease cycles. This dynamic interplay not only optimizes soil health but also stands as a testament to the symbiotic relationship between legumes and their surrounding environment.

C. Biodiversity Enhancement and Ecosystem Services

The impact of legumes reverberates through the intricate web of biodiversity and ecosystem services. These crops, with their blossoming flowers and nutritious pods, become vibrant hubs for a diverse array of pollinators and beneficial insects. From bees to butterflies, legumes contribute to the sustenance of these vital contributors to biodiversity. Beyond this, their roots play a pivotal role in stabilising soil structure, mitigating erosion and acting as guardians against ecosystem degradation. Through their multifaceted roles, legumes transcend the boundaries of mere crops, becoming champions of biodiversity preservation and custodians of essential ecosystem services.

As we uncover the ecological facets of legume cultivation in India, it becomes evident that these plants are not just providers of sustenance but also stewards of environmental harmony, weaving a delicate balance between agriculture and nature.

Agronomic Practices for Optimal Legume Cultivation in India

As we delve into the intricacies of legume cultivation, mastery of agronomic practices becomes paramount for a successful and sustainable harvest. This chapter unravels the essential methodologies tailored to the Indian agricultural landscape, ensuring not only bountiful yields but also the longevity of fertile farmlands (Kebede, 2021).

A. Selection of Suitable Varieties and Strategic Crop Rotation

The art of cultivating legumes commences with the judicious selection of varieties, a decision that shapes the very foundation of crop performance. Local conditions dictate the adaptability of legume varieties, considering factors like maturity duration, disease resistance, environmental resilience and yield potential. Farmers, acting as stewards of the land, must meticulously align these choices with the unique characteristics of their agricultural ecosystem. Further amplifying the wisdom of sustainable cultivation is the practice of strategic crop rotation. Legumes, interwoven with other crops in rotations, contribute to maintaining soil health and orchestrating a defense against pests and diseases. This symbiotic dance not only breaks disease cycles but also enriches the soil with essential nutrients, fostering a harmonious coexistence between crops.

B. Soil Preparation, Sowing and Fertiliser Management: Crafting the Fertile Canvas

The canvas of successful legume cultivation begins with meticulous soil preparation - an art involving deep tillage, precise leveling and adept weed control to sculpt an optimal seedbed. The choice of sowing depth and timing, intricate decisions influenced by the legume species and local climate nuances, adds nuance to this agricultural tapestry. Fertiliser management, a crucial brushstroke in this artistic process, demands a nuanced approach. Soil testing results and the specific needs of legumes guide the application of fertilisers. Remarkably, legumes, with their inherent nitrogen-fixing prowess, necessitate less external nitrogen fertiliser. This eco-friendly trait underscores the sustainability of legume cultivation, aligning with modern agricultural paradigms.

C. Irrigation and Water Management: Nurturing Legumes Through the Thirsty Seasons

Legumes, diverse in their water requirements, demand a delicate dance of irrigation and water management. Some legumes thrive in arid conditions, while others yearn for moist environments. Customized irrigation schedules, tailored to the specific needs of each legume species and responsive to local water availability, become the conductor's wand in this agricultural symphony. Innovative practices, such as drip irrigation, not only conserve water but also enhance overall water use efficiency.

D. Weed Control and Pest Management: Guardians Against Adversaries

The journey of legume cultivation encounters adversaries in the form of weeds and pests. Weeds, vying for resources, can pose a significant threat to yields. The art of timely weed control, whether through manual intervention, mechanical weeding, or judicious herbicide use, emerges as a safeguard for sustained legume productivity. The realm of pest management introduces an integrated approach, orchestrating a symphony of biological, cultural and chemical control methods (Roy *et al.*, 2023). This holistic strategy minimises pest damage, ensuring a balance between crop protection and environmental stewardship.

E. Harvesting and Threshing: Culmination of Agricultural Artistry

As legumes reach their peak of maturity, the art of harvesting takes center stage. Timing becomes paramount, with each legume species presenting specific maturity indicators that guide the harvesting process. Threshing, the final step in this agricultural performance, involves the meticulous separation of grains or seeds from pods or husks. The choice of threshing method - be it manual, machine-driven, or done through combine harvesters - depends on the scale of production and available resources. In this intricate dance of agronomic practices, legume cultivation in India emerges not only as a scientific pursuit but also as an art form, where the careful coordination of choices and practices creates a masterpiece of sustainable and profitable harvests.

Challenges and Hurdles in the Tapestry of Legume Cultivation in India

As we navigate the agricultural tapestry of legume cultivation in India, the journey is not without its challenges and impediments. Despite the myriad benefits they bring, legumes encounter several constraints that cast shadows on their optimal production, requiring thoughtful strategies and interventions (Palai *et al.*, 2019).

A. Abiotic Stresses: Drought, Heat and Salinity

In the vast canvas of India's agricultural landscapes, the brushstrokes of abiotic stresses, particularly drought, heat and salinity, create intricate challenges for legume cultivators. Arid and semi-arid regions, grappling with the relentless grip of drought, pose a significant hurdle to legume cultivation. Heat stress and salinity further complicate matters, impacting not only the growth of legumes but also the quality and yield of the harvest. To surmount these challenges, a pivotal solution lies in the development of legume varieties that demonstrate resilience in the face of adversity. The creation of drought-tolerant, heat-resistant and salinity-tolerant varieties becomes a beacon of hope, offering a pathway to expand legume cultivation into stress-prone areas and bolster the resilience of India's agricultural tapestry.

B. Biotic Stresses: Diseases, Insects and Pests

The delicate dance of legume cultivation encounters a formidable adversary in the form of biotic stresses - diseases, insects and pests. These challenges, capable of inflicting severe yield losses, demand effective management strategies. The arsenal against biotic stresses includes the deployment of resistant varieties, the integration of biological control agents and judicious pesticide applications. Through this multifaceted approach, the impact of diseases, insects and pests can be minimised, allowing legumes to flourish amidst potential threats.

C. Postharvest Losses and Storage Predicaments

The culmination of the agricultural performance brings forth a new set of challenges - postharvest losses and storage issues. The vulnerability during harvesting, drying, storage and transportation stages can significantly dent the profitability of legume cultivation. Inadequate storage facilities and improper handling practices contribute to these losses, acting as hurdles in the path to reaping the full benefits of a harvest. To address this, a spotlight on postharvest management practices becomes imperative. Improving these practices and developing efficient storage facilities stand as critical endeavours, promising not only minimised losses but also enhanced returns for the diligent farmers who steward the legume fields.

D. Market Fluctuations and Price Risks

The final act in the saga of legume cultivation is often marked by the unpredictability of market fluctuations and price risks. This uncertainty poses a challenge for farmers seeking stable incomes from their legume produce. Market interventions, such as price support mechanisms and futures trading, emerge as crucial instruments in the symphony of solutions. By stabilising prices and safeguarding farmers' interests, these interventions provide a safety net against the volatility inherent in the market, fostering a more secure environment for legume cultivators.

In the face of these challenges, the story of legume cultivation in India becomes not just a tale of obstacles but also one of resilience, innovation and strategic interventions. As we navigate through these challenges, the commitment to sustainable and profitable legume harvests stands as a testament to the unwavering spirit of the farmers who till the fields of possibility.

Strategies for Elevating Legume Production and Productivity in India: A Holistic Approach

As we navigate the challenges and complexities inherent in legume cultivation, the journey toward optimal production demands a comprehensive strategy - one that seamlessly integrates technological advancements, policy interventions and the empowerment of farmers through education (Hashim *et al.*, 2023; Pandey *et al.*, 2019).

A. Development of High-Yielding and Stress-Tolerant Varieties: Nurturing Genetic Excellence

At the heart of this strategy lies the imperative need for genetic improvement. Breeding programs, acting as artisans of agriculture, become pivotal in the development of high-yielding, stress-tolerant and disease-resistant legume varieties. These enhanced genetic strains stand as the vanguard against challenges, promising not only increased productivity but also adaptability to the diverse tapestry of agro-climatic conditions that characterise India's agricultural mosaic.

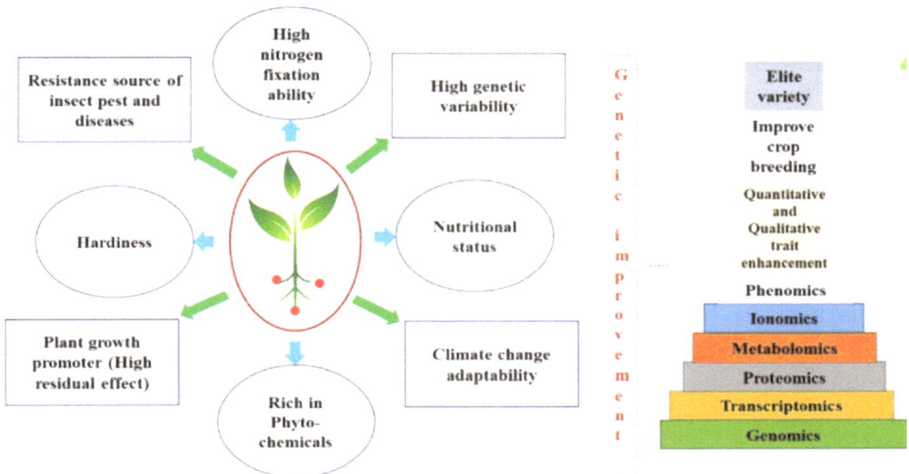

Fig. 5: New technologies to improve legumes by making them more nutritious, resistant to stress and higher-yielding (Samal *et al.*, 2023)

B. Promotion of Sustainable Agricultural Practices: Cultivating Harmony with the Environment

The cultivation of legumes is not just an agricultural pursuit; it is a delicate dance with the environment. Sustainable agricultural practices, akin to the choreography of conservation tillage, integrated pest management and organic farming, emerge as the guiding principles. These practices, designed to promote soil health, enhance water use efficiency and reduce reliance on synthetic inputs, become the pillars of environmentally friendly and sustainable legume cultivation. As farmers embrace these practices, they sow the seeds of a harmonious coexistence between agriculture and nature.

C. Strengthening Extension and Farmer Education: Empowering the Custodians of the Land

The dissemination of knowledge becomes a beacon in the quest for enhanced legume cultivation. Effective extension services and farmer education act as conduits, carrying the wealth of insights about improved varieties, sustainable practices and strategies for managing pests and diseases. By empowering farmers with the requisite knowledge and skills, this strategy aims to not only elevate their understanding but also enhance their cultivation practices and overall productivity in the legume fields.

D. Market Development and Policy Support: Sowing Seeds of Economic Viability

The final steps in this strategic masterpiece involve market development and policy support. Robust market linkages and the promotion of value addition become catalysts in elevating the profitability of legume cultivation. Government policies, serving as pillars of support, come in the form of price support mechanisms, subsidies for improved inputs and initiatives facilitating market access. By fostering an environment that ensures fair returns for the hard work of farmers, these policies play a crucial role in securing the economic viability of legume cultivation.

In this intricate dance of strategies, legume cultivation in India transforms from a mere agricultural endeavor into a holistic pursuit - one that not only overcomes challenges but also nurtures sustainability, empowers farmers and ensures economic viability. As each strategy plays its part, the tapestry of legume cultivation becomes richer, more resilient and a testament to the potential of agriculture as a harmonious partnership between nature and human ingenuity.

Future Horizons of Legume Cultivation in India: Pioneering Sustainability and Innovation

As we gaze into the future of legume cultivation in India, it becomes evident that legumes are poised to emerge as stalwarts in the realm of sustainable agriculture, ushering in a new era marked by environmental consciousness and nutritional fortification (Janila *et al.*, 2016; Okumu *et al.*, 2023).

A. Role of Legumes in Sustainable Agriculture: Guardians of Soil and Nutrient Harmony

Legumes, with their innate ability to harness nitrogen from the air, are destined to assume a pivotal role in steering Indian agriculture toward sustainability. This natural nitrogen fixation prowess holds the key to reducing dependence on synthetic nitrogen fertilisers, mitigating environmental impacts and fostering practices that prioritise sustainable nutrient management. Beyond this, legumes contribute to the improvement of soil health, the sequestration of carbon and the provision of essential ecosystem services, thus weaving themselves seamlessly into the fabric of sustainable agricultural systems (Stagnari *et al.*, 2017).

B. Contribution of Legumes to Nutritional Security: Nourishing the Nation's Health

In the quest for nutritional security, legumes stand as guardians of well-being. Laden with essential nutrients like protein, iron and zinc, they become a cornerstone in addressing malnutrition, especially among vulnerable populations. The promotion of legumes in diets emerges not merely as a culinary choice but as a powerful tool for improved public health outcomes. In a world grappling with rising meat consumption and its environmental implications, legumes present themselves as a sustainable and nutritious alternative, ensuring a pathway to nutritional security. The significant nutritional benefits and health-promoting properties of legumes and pulses include:

- Nutritional Value: Legumes and pulses are rich sources of proteins, complex carbohydrates, dietary fiber, lipids, vitamins and minerals, making them valuable components of a balanced diet.
- Complex Carbohydrates: The abundance of complex carbohydrates in legumes contributes to enhanced satiety, reduced glycemic response and increased fecal bulk, promoting digestive health and aiding in the management of conditions such as diabetes and obesity.
- Bioactive Compounds: Legumes contain phytochemicals such as phenolics, inositols, phytosterols, saponins, phospholipids and γ-aminobutyric acid, which exhibit antioxidative, cardioprotective, anticancer, anti-inflammatory and other health-promoting properties.
- Nutraceutical Properties: Legume proteins exhibit antioxidative, antidiabetic, antihypertensive, anti-hypercholesterolemic and antiproliferative properties, potentially reducing the risk of non-communicable and lifestyle-related diseases such as obesity, type-2 diabetes, strokes/coronary artery diseases and certain cancers.
- Bioactive Proteins/Peptides: Specific small proteins and peptides released upon the hydrolysis of storage proteins in legumes contribute to their nutraceutical properties, including antioxidant, antimicrobial, anti-inflammatory, antihypertensive, cardio-protective, anticancer, antidiabetic and anti-obesity effects.

These properties make legumes and pulses valuable for promoting human health and well-being and their inclusion in diets can contribute to overall nutritional balance and disease prevention (Shevkani, 2023).

C. Potential of Legumes in Value-Added Products: Crafting Economic Prosperity

The economic horizon of legume cultivation holds promise through value addition - a realm where innovation and economic opportunities intertwine. Processing legumes into an array of value-added products, including flours, tofu, tempeh and protein isolates, opens new vistas. This diversification not only expands market reach but also captivates a broader consumer base, ultimately enhancing the income of farmers. The future of legume cultivation becomes a canvas where economic prosperity is painted with the brushstrokes of value-added innovation.

D. Integration of Legumes in Bio-Based Industries: Navigating Toward a Greener Economy

As the world pivots towards a greener economy, legumes step onto the stage as protagonists in bio-based industries. Serving as valuable feedstocks for biofuel production, bioplastics and various other bio-based products, legumes offer sustainable alternatives to traditional fossil fuels. This integration not only aligns with environmental sustainability but also charts a course toward reduced reliance on non-renewable resources, heralding a future where legumes contribute to a more ecologically balanced world. A comprehensive set of strategies aimed at achieving self-sufficiency in pulse production in India encompass short-term, medium-term and long-term approaches, addressing critical aspects such as cropping systems, skill development, insurance mechanisms, seed delivery, procurement, expansion of cultivation areas, farmer organisation, farm equipment development and stress-resistant pulse varieties. By focusing on these strategies, India aims to overcome challenges in pulse production and enhance the nutritional security of its population (Hashim *et al.*, 2023).

In this unfolding narrative of the future, legume cultivation in India transcends its agricultural roots to become a beacon of sustainability, nutritional security, economic prosperity and environmental stewardship. The horizon is ripe with potential and legumes stand poised as agents of positive change, weaving a story where agriculture becomes a harmonious dance between tradition and innovation.

Legumes in Indian Gastronomy: A Symphony of Tradition, Nutrition and Sustainability

India's culinary kaleidoscope is a canvas painted with a rich tapestry of flavours, textures and aromas, where legumes stand not just as ingredients but as culinary luminaries, imparting depth, nutrition and versatility to the diverse array of dishes (Singh *et al.*, 2022).

Table 1: A Culinary Exploration of Legumes: From Humble Bean to Global Delight

Common Name	Food Uses
From Sweet to Savory	
Adzuki beans	These tiny red gems grace Japanese sweets and confections, while also lending their earthy flavor to therapeutic soups.
Anasazi beans	Boiled for a hearty meal, enjoyed as a satisfying snack, or simmered into flavorful soups, Anasazi beans offer comfort and sustenance.
Black gram	A staple of Indian cuisine, black gram transforms into fragrant dhal, ferments into airy idli and crispy dosa and crisps up into thin papad.
Black turtle beans	The heart of Latin American bean soups, black turtle beans add depth and richness to every spoonful.
Black-eyed peas	A versatile ingredient, black-eyed peas are enjoyed boiled as a snack or part of a meal, transformed into savory cakes like akara in West Africa and steamed into the comforting pudding known as moi moi.
Beyond Soups and Stews	
Chickpea	A cornerstone of Middle Eastern and Mediterranean cuisine, chickpeas take center stage in falafel and hummus. They can be boiled, fried, roasted, or crushed into snacks, simmered in dhal and curries, milled into flour for bread, or even fermented into fluffy dhokla.
Faba bean	Enjoyed whole for their simple, satisfying flavour, faba beans are a culinary icon that transcends cultures.
Kidney beans	The backbone of Mexican chili and the most popular legume in America, kidney beans offer a hearty and versatile bean experience.
Lentils	A mainstay of Indian cuisine, lentils are the foundation of countless soups and stews. Their earthy flavor and nutritional profile make them a beloved ingredient across the globe.
From Sprouts to Sweets	
Mung beans/Green gram	These versatile legumes are enjoyed as crunchy bean sprouts, cooked whole in savory dishes, or sweetened into delicious desserts. Their flour finds its way into baked goods and transparent noodles, while patties and sweets add another dimension to their culinary repertoire.
Peanut/Groundnut	From the creamy spread of peanut butter to the crunchy satisfaction of roasted peanuts, this legume has a global reach. Peanut bars and flour offer further uses, while boiled or roasted peanuts remain a timeless snack.
Peas	A familiar sight in soups and dhal, peas offer a simple yet satisfying flavor that complements a variety of dishes.

Soybean	A cornerstone of Asian cuisine, soybeans take center stage in dishes like tofu and natto. They are roasted for snacks, transformed into milk and yoghurt, sprouted for added nutrition and find their way into dishes ranging from curd to yuba.
Beyond the Plate	
Tamarind	The pulp of this tart fruit is used in food and beverages, while its flour acts as a soup thickener. It even finds use as a remedy for diarrhea and constipation.

Conclusion

In India's rich narrative of agriculture and cuisine, legumes play a crucial role in contributing to sustainability, nutrition and economic prosperity. To unlock their potential, addressing challenges, promoting sustainable practices and enhancing post-harvest management are essential. Legumes, with their nitrogen-fixing ability, are integral to sustainable agriculture, minimising environmental impact. They also combat malnutrition, offer a sustainable protein alternative and contribute to value-added products, boosting economic opportunities. As feedstocks for bio-based industries, legumes pave the way for a greener economy. Embracing legumes offers India a path to a brighter and sustainable future, aligning with global challenges and celebrating culinary heritage.

References

Chaturvedi, S. K., and Sandhu, J.S. (2020). Silent revolution in pulses production – India marching towards self-sufficiency. The Indian Journal of Agricultural Sciences, 90(1), 17-24. DOI:10.56093/ijas.v90i1.98521

Das, A., and Ghosh, P.K. (2012). Role of Legumes in Sustainable Agriculture and Food Security: An Indian Perspective. Outlook on Agriculture, 41(4), 279-284. DOI: 10.5367/oa.2012.0109

Hashim, M., Kumar, N., Singh, R., Nath, C., and Hazra, K. (2023). Steps towards self-sufficiency in pulse production in India. Indian Farmer, 10(3), 130-136.

Janila, P., Rupavatharam, S., Kumar, C.S., Samineni, S., Gaur, P., and Varshney, R. (2016). Technologies for intensification of production and uses of grain legumes for nutrition security. Proceedings of the Indian National Science Academy, 82, 1541-1553.

Kebede, E. (2021). Contribution, Utilisation and Improvement of Legumes-Driven Biological Nitrogen Fixation in Agricultural Systems. Frontiers in Sustainable Food Systems, 5. DOI: 10.3389/fsufs.2021.767998

Okumu, O., Otieno, H.M.O, and Okeyo, G.O. (2023). Production systems and contributions of grain legumes to soil health and sustainable agriculture: A review. Archives of Agriculture and Environmental Science, 8(2), 259-267. DOI: 10.26832/24566632.2023.0802024

Palai, J.B., Jena, J. and Maitra, S. (2019). Prospects of underutilised food legumes in sustaining pulse needs in India-A review. Crop Research, 54.

Pandey, V.L., Dev, S.M., and Mishra, R. (2019). Pulses in eastern India: production barriers and consumption coping strategies. Food Security, 11(3), 609-622. DOI: 10.1007/s12571-019-00917-y

Roy, D. K., Ranjan, S., and Sow, S. (2023). Production potential and economics of integrated weed control measures in ginger. Indian Journal of Weed Science, 55(3), 328-332. DOI: 10.5958/0974-8164.2023.00061.8

Samal, I., Bhoi, T. K., Raj, M. N., Majhi, P. K., Murmu, S., Pradhan, A. K., et al. (2023). Underutilised legumes: nutrient status and advanced breeding approaches for qualitative and quantitative enhancement. Frontiers in Nutrition, 10. DOI: 10.3389/fnut.2023.1110750

Sarkar, S., Panda, S., Yadav, K., and Kandasamy, P. (2020). Pigeon pea (Cajanus cajan) an important food legume in Indian scenario-A review. Legume research-an international journal, 43(5), 601-610.

Sharma, V., and Shukla, V. (2014). Scenario of Lentil in India–A review. Advances in life sciences, 3(1), 01-06.

Shevkani, K. (2023). Protein from land—legumes and pulses Future Proteins (pp. 35-68): Elsevier.

Singh, N., Jain, P., Ujinwal, M., and Langyan, S. (2022). Escalate protein plates from legumes for sustainable human nutrition. Frontiers in Nutrition 9. DOI: 10.3389/fnut.2022.977986

Sow, S., and Ranjan, S. (2021). Popularization of Pulse Production for Ensuring Nutritional and Food Security. Agriculture & Food: E-Newsletter, 3(2), 16-19.

Stagnari, F., Maggio, A., Galieni, A., and Pisante, M. (2017). Multiple benefits of legumes for agriculture sustainability: an overview. Chemical and Biological Technologies in Agriculture, 4(1), 2. DOI: 10.1186/s40538-016-0085-1

Thilakarathna, M. S., McElroy, M. S., Chapagain, T., Papadopoulos, Y. A., and Raizada, M. N. (2016). Belowground nitrogen transfer from legumes to non-legumes under managed herbaceous cropping systems. A review. Agronomy for Sustainable Development, 36(4), 58. DOI: 10.1007/s13593-016-0396-4

9

Benefits of Millets Cultivation in India

Nirmul Kumar[1], Aditya Bharti[2] and Kumari Sristi[3]

[1]*Department of Plant Pathology and Nematology, Dr. Rajendra Prasad Central Agricultural University, Pusa-848125, Bihar*

[2]*Department of Plant Pathology, Bihar Agricultural University Sabour Bhagalpur-813210, Bihar*

[3]*Department of Agronomy, Bihar Agricultural University Sabour, Bhagalpur-813210 Bihar*

Abstract

Millet, a nutritious cereal, is a food with high energy content that was domesticated and cultivated 10,000 years ago. It includes jowar, bajra, ragi, and small millets. Small millets include Barnyard millet, proso millet, kodo millet, little millet, and foxtail millet. Millet cultivation is often carried out on degraded and marginal lands that receive little rainfall and are poor in nutrients. Over-reliance on cereals after the Green Revolution and people's current sedentary lifestyle have increased health-related disorders such as obesity, diabetes, coronary heart disease, digestive disorders, and the risk of colon, breast, and esophageal cancer. The only way to fight these diseases is to include nutritious millets in our daily diet. Millets are unique due to their richness in fiber, antioxidants, minerals, calcium, phytochemicals, polyphenols and proteins, making them a solution to combat health-related disorders. Recent global climate change has reduced the productivity of major cereals and paved the way for the integration of millets into agricultural production systems to create climate-resilient cropping systems. Millet is a C_4 crop with higher photosynthetic efficiency, shorter duration, higher dry matter production, and high heat and drought tolerance. Keeping in mind the above advantages of millet, efforts have been stepped up in collecting, preserving, and using millet genetic material in breeding programmes. Recently, several private and government agencies have ventured into value addition of millets for the production of food and non-food products. However, governments play an important role in formulating policies to promote the cultivation and consumption of millets.

***Keywords**: Antioxidant, Climate-resilient, Millets, Nutritious, Tolerance*

Introduction

Millet is a small grain cereal grass bearing coarse seeds, and since ancient times has been considered a valuable and nutritious grain. It includes jowar, bajra, ragi and small millets. Small millets include Barnyard millet, proso millet, kodo millet, little millet and foxtail millet. It is rich in fiber, protein, iron, calcium, magnesium, zinc, vitamin B, has a low glycemic index and has antioxidant, anti-inflammatory and anti-allergic properties (Ranjan *et al.*, 2023). These have been added to the modern lifestyle as healthy foods because they are beneficial for human health due to the presence of nutritional ingredients. The nutritional quality of food is the most important parameter for maintaining human health and comprehensive physical health. Good nutrition is the driving force for growth and maximisation of human genetic potential (Radhila *et al*, 2011). Food quality in the diet must be considered to maintain optimal human health and fitness to address the deep-seated problem of malnutrition. Diversification of food production at both the national and household levels should be encouraged, in addition to increasing domestic productivity and technology (Singh and Raghuvanshi, 2016). Some agricultural foods are not used by humans as staple foods due to human ignorance. Millet is one of them. Millet is used as food for animals and birds. Millet has many nutritional and medical functions (Yang *et al*, 2012). These are underutilised and neglected crops due to lack of knowledge among people and some important issues such as cooking quality, taste and low bioavailability of millets. These problems can be solved to make them a valuable source of food for poor families to combat malnutrition and become an important source of income. Millet is a very important crop with the following characteristics: Millet is known to be drought-resistant, pest-resistant, and has a short growing period compared to other major cereals (Devi *et al*, 2011). Due to the superior features mentioned above;, millet is receiving particular attention in developing countries such as India, China and some countries in Africa for its use as food. Some developed countries are also particularly interested in millets due to their good potential in bioethanol and biofilm production (Lei and Jacobsen, 2004). Millet is an important food crop in developing countries as it contains major and minor nutrients in significant amounts.

FAO reports that traditional food processing methods (such as peeling, milling, sprouting, fermenting, malting, roasting, etc.) are often used to process millet food products to improve their properties such as edibility, nutrition, and sensory properties. However negative changes in millet are inevitable because industrial processing methods are not yet as developed compared to other cereals (FAO, 2012). Millets can be a source of healthy value-added food products in a variety of varieties for both traditional and non-traditional millet users (Mal *et al*, 2010). India, with its vast and diverse agriculture, has long depended on a variety of crops to meet the nutritional needs of its people. In recent years, there has been a renewed focus on millet farming, recognising its central role in addressing the multifaceted challenges facing the country. Millet cultivation in India is particularly important due to its resilience to adverse climatic conditions, nutritional richness, cultural significance, and potential to contribute to sustainable agriculture and economic growth. As the country grapples with problems ranging from the impacts of climate change to malnutrition, replanting millets has emerged as a strategic and holistic solution, offering a roadmap towards food security, environmental sustainability and preservation of traditional agricultural practices. This transition to millets not only reflects an agricultural renaissance but also signifies a commitment to building a diverse and resilient food system for the well-being of current and future generations in India.

Production of Millets

Millet cultivation in India has seen a resurgence in recent years. The area devoted to millet cultivation has gradually increased as farmers and policymakers recognise the benefits of these resilient crops. States like Karnataka, Rajasthan, Andhra Pradesh, Telangana, Maharashtra and Tamil Nadu are major contributors to millet cultivation. India is one of the world's largest millet producers. Total millet production, which includes varieties such as pearl millet (bajra), finger millet (ragi), foxtail millet and sorghum (jowar), shows a positive trend. Production figures fluctuate depending on climatic conditions and agricultural practices.

The world's largest millet producer is India. In 2020, the two types of millets grown in India, pearl millet (Bajra) and sorghum (Jowar), accounted for about 19% of global production. In 2020, India produced 40.51% of the world's millet production, followed by 8.09% of millet production by sorghum. India's primary millet-producing states include Rajasthan, Karnataka, Maharashtra, Uttar Pradesh, Haryana, Gujarat, Madhya Pradesh, Tamil Nadu, Andhra Pradesh, and Uttarakhand. Currently, these ten states account for nearly 98% of India's millet production in 2020-21 (MOA&FW GOI). Millet production is spread across six states – Rajasthan, Karnataka, Maharashtra, Uttar Pradesh, Haryana, and Gujarat – accounting for over 83% of the total production. 28% of India's total millet production comes from Rajasthan. India produces many different types of millets, including pearl millet, sorghum, finger millet, foxtail millet, kodo millet, Barnyard millet, proso millet, small millet, and imitation millets such as buckwheat and amaranth. Three types of millets account for the majority of India's total millet production: pearl millet (Bajra), sorghum (Jowar) and finger millet (Ragi) (Sukumaran *et al.*, 2023). In 2020-21, total millet production in India was around 18.02 million tonnes (MOA&FW). This includes many different types of millets, with pearl millet and sorghum being among the main varieties.

Government Initiatives

The Indian government actively encourages millet cultivation through various programmes and initiatives. The National Mission for Sustainable Agriculture (NMSA) and National Food Security Mission (NFSM) are among the programmes aimed at increasing millet production and improving farmers' income. The Indian Institute of Millets Research (IIMR) has upgraded recipes originally developed by the Directorate of Sorghum Research (DSR), a central body under ICAR dedicated to sorghum, to encompass all millets through extensive research and development (Dayal *et al.*, 2023). The NAIP-Millets Value Chain sub-project is particularly focused on enhancing selected millet foods, with a special emphasis on sorghum. If the model proves successful, it has the potential to be replicated for other millets, providing significant benefits to consumers, households, and entrepreneurs. These efforts aim to encourage the preparation of various value-added products with health benefits that may not be readily available in the market. The promotion of value-added products by households, entrepreneurs, self-help groups, and businesses, both small and large, has the potential to enhance income levels. Moreover, these products can contribute to improving consumers' health and socioeconomic status. Collaborative endeavours have led to the development and trial commercialisation of nutritionally dense and convenient sorghum products in Hyderabad. Scientific investigations have been conducted on dishes like idli, dosa, and others, incorporating sorghum and various millets such as rice and wheat. Effective promotion and policy advocacy have successfully persuaded entrepreneurs and policymakers to prioritise the utilisation of sorghum in diverse culinary applications.

Types of Millets

Millets vary in shape, plant type, grain type, maturity, morphological characteristics, etc. Millet is divided into two types: large millet and small millet. The main types of millets are pearl millet (*Pennisetum glaucum*), the most widely used for human consumption (ICRISAT, 2007), foxtail millet (*Setaria italica*), common or white millet (*Panicum miliaceum*) and finger millet (*Eleusine coracana*). Small millets include Barnyard millet (*Echinochloa spp.*), kodo millet (*Paspalum scrobiculatum*), small millet

(*Panicum sumatrense*), guinea millet (*Brachiaria deflexa*), browntop millet (*Urochloa ramosa*), Teff (*Eragrostis tef*), Fonio (*Digitaria exilis*), sorghum (*Sorghum spp.*) and Job's Tears (*Coix lacryma-jobi*) (Adekunle, 2012).

Fig. 1: Different types of millets

Millets and Its Potential

Millet is a small, round grain that belongs to the small-seeded grass family (Poaceae). It is characterised by its outstanding ability to survive in less fertile soils, drought resistance, pest resistance, short growing season, and can be cultivated year-round and worldwide. The word millet originates from the beginning of human civilisation, millet is considered the first domesticated grain (Shahidi and Chandrasekara, 2013). India is the largest producer of millets (Bhattacharjee, 2007). Millet seeds are especially rich in protein, calcium, fiber and polyphenols. Millets are known to contain significant amounts of sulphur-containing essential amino acids such as methionine and cysteine (Obilana and Manyasa, 2002).

Nutrient Dense Millets

Millets contain 60-70% carbohydrates, 7-11% protein, 1.5-5% fat and 2-7% crude fiber and are also rich in vitamins and minerals. They are an excellent source of vitamin B, magnesium, and antioxidants. Millets are also a good source of other nutritional minerals such as manganese, phosphorus, and iron. Millet protein is a good source of essential amino acids, except lysine and threonine, but is relatively rich in the sulphur amino acids methionine and cysteine (Singh *et al*, 2012). In addition, some essential fatty acids such as linoleic, oleic, and palmitic acids exist in free form, and mono galactosyl, diacylglycerol, di galactosyl diacylglycerols, phosphatidylethanolamine, phosphatidylserine and phosphatidylcholine in the bound form are also present in millet (Bagdi *et al*, 2011). Other fatty acids, specifically arachidic acid, behenic acid and erucic acid, are found in trace amounts. Millet oil can be a good source of linoleic acid and tocopherols (Amadou *et al*, 2011). Millet is a gluten-free alkaline grain (Moreno *et al.*, 2014). B vitamins such as niacin, folacin, riboflavin, thiamine, and phosphorus are present in millet and play a major role in the body's energy synthesis.

Tabl 1. Nutritional benefits of millets (For 100g of each cereal)

Millets	Protein (g)	Fiber (g)	Mineral (g)	Iron (mg)	Calcium (mg)
Pearl millet	10.6	1.3	2.3	16.9	38
Finger millet	7.3	3.6	2.7	3.9	344
Foxtail millet	12.3	8.0	3.3	2.8	31
Proso millet	12.5	2.2	1.9	0.8	14

Importance of Millet Cultivation in India

Millet cultivation is of significant importance in India for many reasons, contributing to the agricultural, economic, nutritional, and environmental aspects of the country. Here are some key reasons that highlight the importance of millet cultivation in India:

1. Drought Resistance: Millet is a hardy crop that can thrive in semi-arid and drought-prone regions where other crops may struggle. Given India's diverse climatic conditions, millets offer a resilient option for farmers, especially in areas with unreliable rainfall.
2. Nutritional value: Millets are very nutritious and rich in essential nutrients like iron, calcium, fiber and antioxidants. Their consumption has been linked to improved health outcomes, making them an important part of combating malnutrition and ensuring food security, especially in areas with limited accessto diverse food sources.
3. Cultural Heritage: Millets have been an integral part of the traditional diet of various regions in India for centuries. They are often deeply rooted in the cultural and culinary heritage of many communities. Promoting

millet cultivation helps preserve traditional agricultural practices and dietary habits.

4. Climate-smart agriculture: Millet cultivation is consistent with the principles of climate-smart agriculture. These crops have lower water and input requirements than many other staple grains. This makes millet ecologically sustainable and contributes to climate change mitigation and adaptation efforts.
5. Income diversification: Millet farming allows farmers to diversify their income. Growing mill*et al*ongside other crops can enhance farmers' resilience to market fluctuations and economic uncertainties, thereby improving their income and overall livelihood.
6. Soil health: Millet is known for its ability to adapt to different soil types and can thrive in less fertile lands. They have a positive impact on soil health by improving soil structure and fertility, while reducing the need for excessive fertiliser use.
7. Biodiversity conservation: Promoting millet cultivation contributes to the conservation of agricultural biodiversity. Crop diversification helps maintain a rich gene pool, which is important for the long-term sustainability of agriculture.
8. Food safety: Millet can play an important role in improving food security, especially in areas at risk of monoculture or dependence on a few crops. Millet's resilience to varying environmental conditions makes it a reliable food source in difficult agricultural areas.
9. Export potential: With growing global interest in healthy and sustainable food, Indian millets have the potential to become an export commodity. This can open up new markets and economic opportunities for farmers, thereby contributing to boosting the country's agricultural exports.

Historical Importance of Millet Crops in Indian Agriculture

Millet crops, integral to Indian agriculture and culture, have a rich history, serving as staples for centuries. Especially in rural and arid regions, they ensure food and nutritional security. Historically, millets were crucial in native farming, prized for their resilience in varied conditions. Consumed as whole grains, flour, malted beverages, and fermented dishes, they enriched diets and protected against weather uncertainties. Millet's adaptability and ability to yield well in challenging circumstances make them a dependable food source for small-scale farmers, contributing significantly to food security. (Tiwari *et al.*, 2023). Despite the Green Revolution's influence in diminishing millet production, there is a renewed enthusiasm for reviving millet crops.

Recent efforts are underway to promote their cultivation and consumption in India, driven by their nutritional benefits, resilience to varying climates, and potential for sustainable agriculture. The historical importance of millets in Indian agriculture provides a strong basis for acknowledging their potential as sustainable crops, underscoring the importance of their revival and promotion of the country's agricultural sustainability.

Environmental Benefits of Millet Cultivation

- Drought Tolerance and Water Use Efficiency: Cultivating millet offers significant environmental benefits, primarily due to its exceptional drought resistance and water-saving capabilities. Tailored to thrive in arid conditions, millet crops are well-suited for cultivation in regions with limited water resources. Unlike major cereal crops such as rice and wheat, millets demand minimal water, showcasing evolved water-saving techniques and adaptations to water scarcity. The thick root systems of millet plants reduce reliance on surface water supplies by enabling them to extract water from deeper soil layers. Millets' ability to maintain yields with minimal irrigation or in rainfed conditions contributes to water conservation and alleviates pressure on water resources. Moreover, millet cultivation plays a vital role in enhancing soil health. The fibrous roots of millet plants improve soil structure and increase organic matter content. The extensive root system aids in soil agglomeration, preventing erosion and enhancing the soil's water retention capacity. In essence, growing millet proves beneficial not only for water conservation but also for promoting soil fertility and health.
- Biodiversity Preservation: Millet crop farming supports biodiversity in agricultureby preserving traditional seed variants and agroecosystem diversity. It safeguards landraces adapted to specific conditions, enhancing genetic diversity and resilience against pests and climate change. Millet cultivation requires fewer chemical inputs, as the crops naturally resist pests and diseases, reducing the need for synthetic pesticides. This eco-friendly approach mitigates the environmental and health impacts of excessive chemical use, preventing soil erosion and water pollution. The environmental benefits of millet production, including drought resistance, efficient water use, soil health enhancement, and biodiversity preservation, make it a sustainable option for agriculture. Promoting millet farming can contribute to overall environmental sustainability in India's agriculture industry.

Economic and Social Benefits

- Economic Viability of Millet Cultivation: In agricultural settings, particularly those with limited resources, cultivating millets provides economic viability and the potential for income generation for farmers. Compared to major cereal crops, millets are cost-effective to grow as they require fewer inputs such as water, fertiliser, and pesticides. This reduction in production costs enhances farmers' profitability. Millets also have a shorter growing season than crops like wheat or rice, allowing farmers to undertake multiple cropping cycles throughout the year, enabling diversification and expanding income options. Furthermore, the increasing consumer interest in the nutritional and medicinal benefits of millets has elevated their demand in both domestic and global markets.
- Market Possibilities and Value Chains: The revelation of millets' nutritional value and their potential for sustainable agriculture has opened up new commercial opportunities. Millets are increasingly incorporated into various foods, including breakfast cereals, snacks, baked goods, and gluten-free products. This diversification in the market not only adds value to millet-based products but also contributes to higher market pricing. Prioritising the strengthening of millet value chains and facilitating farmer access to markets is a key focus. Government initiatives, farmer-producer organisations, and non-governmental organisations are actively involved in enhancing millet processing, storage, and marketing infrastructure. These efforts aim to empower farmers with improved access to larger markets, enabling them to generate income at fair rates.
- Improvement of Small-Scale Farmer's Livelihoods: Millet production offers significant advantages to small-scale farmers who often face challenges such as climate variability, resource constraints, and market limitations. Millets, requiring minimal input and displaying resilience to environmental challenges, emerge as a favourable choice for small-scale farming systems. Their cultivation enables farmers to diversify their planting systems, leading to increased income and food security. Millets serve as a reliable safety net for small-scale farmers, especially in the face of droughts and other climate uncertainties, providing stability during crop failures. The shorter cultivation period of millets further allows farmers to allocate more time and resources to other income-generating activities, contributing to an overall improvement in their standard of living.

- Women's Empowerment and Nutritional Security: Millet farming is of great significance for women's empowerment and nutritional security. Often tasked with managing nutrition and food preparation in rural households, women can significantly enhance household nutrition through millet cultivation. Millets, with their high nutritional content and versatile culinary applications, offer an opportunity to improve nutritional outcomes. Empowering women is a key outcome of millet farming, providing them with opportunities for income generation and active participation in decision-making processes. Women often play vital roles in the cultivation, processing, and marketing of millets. Recognising and promoting women's participation in millet value chains can contribute to their economic and social empowerment. Millets, rich in micronutrients, offer a solution to address malnutrition and food security issues by diversifying diets. Additionally, millets provide affordable and nourishing food options for vulnerable communities, enhancing nutrition security at both the family and community levels.

The economic and social advantages of millet cultivation, including its economic viability, market prospects, livelihood improvement for small-scale farmers, and women's empowerment, highlight millet's potential as a sustainable agricultural option. Promoting millet cultivation in India not only facilitates economic growth but also addresses important social issues such as nutrition security, gender equality, and livelihood improvement. (Dwivedi *et al.*, 2023).

Future Perspectives

In the future perspective of millet development, the focus is on increasing consumer demand through innovative campaigns and outreach efforts, emphasising the nutritional worth and health advantages of millets. The strategy involves showcasing their adaptability and incorporation into a variety of food products, aiming to make millet-based products more appealing to consumers. Simultaneously, there is a commitment to invest in post-harvest infrastructure, storage facilities, processing units, and packaging facilities. This comprehensive approach is designed to preserve millet grain quality, promote value addition, and contribute to the growth of a diverse millet-based product market. The establishment of effective supply chains and the strengthening of market ties are prioritised to connect millet producers to larger markets, ensuring fair prices and income generation within the sector (Dwivedi *et al.*, 2023).

Capacity building and knowledge dissemination are essential components of the future plan. Training sessions, workshops, and field trips will be organised to provide farmers, extension agents, and policymakers with the necessary information and skills for successful millet farming. The strategy also includes recording and disseminating traditional millet farming knowledge and techniques to preserve native millet types and encourage their cultivation.

Conclusion

Millet cultivation in India holds pivotal importance in guaranteeing food security, safeguarding cultural heritage, fostering sustainable agriculture, and enhancing farmers' livelihoods. Supporting and recognising the cultivation of millets can have profound positive impacts on the agricultural landscape and the well-being of communities throughout the nation. In India, where the majority of farmers are small and marginal, facing numerous challenges in farm management, it is imperative for farming practices to be more resilient and diversified to meet the country's food demands while ensuring the sustainable use of natural resources. Although it took a considerable amount of time, the government is now actively endorsing millet farming, especially in urban areas. The growing attention towards millets, coupled with the designation of 2023 as the International Year of Millet, enhances their global significance in terms of promotion and cultivation.

References

Adekunle, A. A., Ellis-Jones, J., Ajibefun, I., Nyikal, R. A., Bangali, S., Fatunbi, A. O., & Angé, A. (2013, June). Agricultural innovation in sub-Saharan Africa: Experiences from multiple stakeholder approaches. Accra, Ghana: Forum for Agricultural Research in Africa (FARA).

Amadou, I., Amza, T., Shi, Y. H., & Le, G. W. (2011). Chemical analysis and antioxidant properties of foxtail millet bran extracts. Songklanakarin Journal of Science & Technology, 33(5).

Amadou, I., Gounga, M. E., & Le, G. W. (2013). Millets: Nutritional composition, some health benefits and processing-A review. Emirates Journal of Food and Agriculture, 501-508.

Bagdi, A., Balázs, G., Schmidt, J., Szatmári, M., Schoenlechner, R., Berghofer, E., & Tömösközia, S. J. A. A. (2011). Protein characterisation and nutrient composition of Hungarian proso millet varieties and the effect of decortication. Acta Alimentaria, 40(1), 128-141.

Balkrishna, A., Shankar, R., Prajapati, U. B., Joshi, R. A., Srivastava, A., & Awasthi, C. (2023). Exploring the common millets of India: A comprehensive review. Journal of Drug Research in Ayurvedic Sciences, 8(Suppl 1), S26-S33.

Bhattacharjee, R., Khairwal, I. S., Bramel, P. J., & Reddy, K. N. (2007). Establishment of a pearl millet [Pennisetum glaucum (L.) R. Br.] core collection based on geographical distribution and quantitative traits. Euphytica, 155, 35-45.

Dayal, P., Prajapati, S., Sow, S., Shukla, V.K., & Ranjan, S. (2023). Millets: Nutritional Profile and Health Benefits as Nutri Cereals. In: Soil and Crop Management Practices for Sustainable Agriculture, Elite Publishing House, pp. 106-118.

Devi, P. B., Vijayabharathi, R., Sathyabama, S., Malleshi, N. G., & Priyadarisini, V. B. (2014). Health benefits of finger millet (Eleusine coracana L.) polyphenols and dietary fiber: a review. Journal of Food Science and Technology, 51, 1021-1040.

Dwivedi, N., Rathore, V., & Sharma, K. (2023). A Review of Millet Crops for Agricultural Sustainability in India. Asian Journal of Agricultural Extension, Economics & Sociology, 41(10), 216-224.

Lei, V., & Jakobsen, M. (2004). Microbiological characterisation and probiotic potential of koko and koko sour water, African spontaneously fermented millet porridge and drink. Journal of Applied Microbiology, 96(2), 384-397.

Mal, B., Padulosi, S., & Bala Ravi, S. (2010). Minor millets in South Asia: learnings from IFAD-NUS Project in India and Nepal.

Moreno Amador, M. D. L., Comino Montilla, I. M., & Sousa Martín, C. (2014). Alternative grains as potential raw material for gluten–free food development in the diet of celiac and gluten–sensitive patients. Austin Journal of Nutrition and Metabolism, 2 (3), 1-9.

Pragya, S., & Rita, S. R. (2012). Finger millet for food and nutritional security. African Journal of Food Science, 6(4), 77-84.

Radhika, G., Sathya, R. M., Ganesan, A., Saroja, R., Vijayalakshmi, P., Sudha, V., & Mohan, V. (2011). Dietary profile of urban adult population in South India in the context of chronic disease epidemiology (CURES–68). Public health nutrition, 14(4), 591-598.

Ranjan, S., Sow, S., Ghosh, M., Padhan, S.R., Kumar, S., Gitari, H., Mirriam, A., & Nath, D. (2023) Nutraceutical properties andsecondary metabolites of quinoa (Chenopodium quinoa Willd.): a review. International Journalof Food Properties, 26(2), 3477-34

Sarita, E. S., & Singh, E. (2016). Potential of millets: nutrients composition and health benefits. Journal of Scientific and Innovative Research, 5(2), 46-50.

Shahidi, F., & Chandrasekara, A. (2013). Millet grain phenolics and their role in disease risk reduction and health promotion: A review. Journal of Functional Foods, 5(2), 570-581.

Singh, K. P., Mishra, A., & Mishra, H. N. (2012). Fuzzy analysis of sensory attributes of bread prepared from millet-based composite flours. LWT-Food Science and Technology, 48(2), 276-282.

Sukumaran Sreekala, A. D., Anbukkani, P., Singh, A., Dayakar Rao, B., & Jha, G. K. (2023). Millet Production and Consumption in India: Where Do We Stand and Where Do We Go?. National Academy Science Letters, 46(1), 65-70.

Tiwari, H., Naresh, R. K., & Pandey, M. (2023). Millets: Key to food and nutritional security in India. Current Agriculture Trends: e-Newsletter. 2023b, 2(2), 1-4.

Yang, X., Wan, Z., Perry, L., Lu, H., Wang, Q., Zhao, C., & Ge, Q. (2012). Early millet use in northern China. Proceedings of the National Academy of Sciences, 109(10), 3726-3730.

10

Nematode Associated Molecular Pattern (NAMP) Mediated Immune Response Against Plant Parasitic Nematodes

Shandeep S.G.[1], Arun A.[1], Haran M.S.R.[2], Deeikshana T.[1] and Vijay S.[3]

[1]*Department of Nematology, Tamil Nadu Agricultural University, Coimbatore, India*

[2]*Department of Agricultural Entomology, Tamil Nadu Agricultural University Coimbatore, India*

[3]*Silkworm Seed Production Centre, National Silkworm Seed Organisation Central Silk Board, Bhawanipur, West Bengal, India*

Abstract

Plant parasitic nematodes like Meloidogyne spp., Globodera rostochinensis, Heterodera spp., Pratylenchus spp., and Radopholus similis, either alone or in association with fungal pathogens cause severe yield and economic losses to the farming community. Globally, fluopyram and fluensulfone are commercially available for the management of nematode complexes. However, plant parasitic nematodes with the help of their esophageal gland proteins which are termed as Nematode Associated Molecular Patterns (NAMP), are known to trigger innate immunity in plants against invading parasitic nematodes. Plants, through their localised receptors such as Pattern Recognition Receptors identify the invading nematodes and trigger innate immunity to prevent further attacks. The plant's innate immune reaction against plant parasitic nematodes is briefly discussed in this chapter.

Keywords: *Nematode Associated Molecular Pattern, Plant innate immunity Parasitic nematodes, Resistance mechanism*

Introduction

Plant Parasitic Nematodes (PPN) belonging to the kingdom Animalia are considered the most serious threat to agriculture and horticulture crops worldwide. The estimated yield loss on average is about 173 billion USD (12.3%), among which developing countries are more prone to nematode infestation (14.46%) than developed countries (8.8%). Nevertheless, these

PPN aggravate the loss by forming disease complexes with other soil-borne pathogens (Jones *et al.*, 2013). PPN gain entry onto the host roots through their secretome which secretes various kinds of proteins that are mandatory for the expression of disease symptoms in host plants. These proteins are generally called Nematode-Associated Molecular Patterns (NAMPs). The management of nematodes through synthetic nematicides paves the way for the production of commodities with toxic residues that may have severe ill effects after consumption. However, the use of resistant plants is another strategy to combat major PPN and suppress their population in soil and roots. Recent research progress has revealed the various defence mechanisms exhibited by plants against the invasion of PPN. Here we will discuss in brief about the various mechanisms followed by the host plants to recognise the invading PPNs and the resulting host immune response in response to recognition of various effector proteins of PPN.

Recognition of PPN

The plants have various kinds of receptor sites for recognising the effector protein produced by PPN, pathogens (bacteria, fungi, viruses). This recognition system is mediated by perceiving the signals either mediated by Pathogen-Associated Molecular Pattern (PAMP) or Damage-Associated Molecular Pattern (DAMP) which are released as a result of damage of host cells by insect pest invasion. The PAMP and DAMP molecules are recognised by the receptor site of the host i.e., Pattern Recognition Receptor (PRR). This recognition leads to triggering of some kind of basal defence mechanism in plants. This kind of induction of the first line of defence mechanism in plants is termed as Pattern Triggered Immunity (PTI) (Boutrot and Zipfel, 2017; Hou *et al.*, 2019).

As a counter mechanism the nematodes secrete effector proteins that interfere with the host's PTI and lead to successful parasitism. The resistant plants recognise the effector proteins through Nucleotide-binding leucine-rich repeat (NB-LRR) receptors which trigger NLR triggered immunity (NLR-TI). These NLR proteins have two subunits, either Toll-interleukin receptor (TIR-1-NLR) or Coiled-coil (CC-NLR) (Jones and Dangal, 2006).

The effector proteins of PPN are found to induce defence responses in plants. The nematode pheromone Ascaroside is the first and only nematode PAMP identified and characterized so far (Manosalva *et al.*, 2015). This ascaroside is found to induce various defence responses in plants such as mitogen-activated protein kinase (MAPK) activation, defence gene expression, and activation of signaling pathways such as the Salicylic acid and Jasmonic Acid pathways.

This compound has been proven to be effective against Meloidogyne sp., and Cyst nematodes and also found to be recognised by various commercial crops such as tomato, potato and cereals. The PRR for recognising the ascaroside was not yet found.

In case of PRR inducing immunity against PPN, leucine-rich repeat (LRR)-RLK encoded by Arabidopsis Nilr1 (nematode-induced LRR-RLK (Mendy *et al.*, 2017). This receptor was found to recognise major PPN such as H. schactii and M. incognita juveniles. Importantly, root knot nematodes RKNs and cyst nematodes CNs have multiple virulence effectors that are able to suppress PTI responses (Chen *et al.*, 2016; Jaouannet *et al.*, 2013; Lin *et al.* 2016; Chen *et al.*, 2018; Naalden *et al.*, 2018; Kud *et al.*, 2019; Yang *et al.* 2019).

Incase of migration of PPN within the host leading to damage of host cells, which produces DAMP molecules. These molecules trigger the immune system of plants against PPN. Upon migration of cyst nematodes, the host cell wall releases oligogalacturonides (OGs). Fungal pathogens produce Polygalacturonase (PG) to digest cell wall materials (D-Ovitdio *et al.*, 2004) and host plants have PG inhibitor proteins (PGIP) which stops the further degradation of host pectin by releasing OG. This OG has been proven to trigger PTI in plants. Arabidopsis has two kinds of PGIP viz., PGIP1 and PGIP2 which are expressed during the invasion of cyst nematodes. Among the two PGIPs, PGIP 1 was found to activate the plant camalexin and indole-glucosinolate pathways to suppress the cyst nematode population.

Cyst nematodes induce the host plants to produce ethylene, delays the formation of feeding sites of PPN (Marhavy *et al.*, 2019). NLR proteins play a crucial role in determining the resistant factors against PPN. This kind of recognition by NLR is encoded through resistance genes (R genes). The secreted effector protein of G.pallida, Gp-RBP-1 induces the hypersensitive reaction (HR) in the presence of Gpa2 (R gene). There are various surface-localised PRRs available to recognise the PPN. Apoplastic Papain like Cysteine Protease (PLCP) RCR3pim fails in tomato to recognise the Venom allergen like proteins of G. rostrochinensis (GrVAP-1). Nevertheless, Cf-2, an extracellular LRR, helps to trigger the HR response during the interaction between GrVAP-1 and (PLCP) RCR3pim (Lozano-Torres *et al.*, 2012). Thus the host plants have various mechanisms to recognise the invading PPNs.

Production of nematicidal compounds in host apoplast

Nematicidal compounds are substances that directly inhibit or arrest the activity of nematodes. One such enzyme is chitinase. This chitinase has the potential to inhibit the activity of a wide range of pathogens including nematodes and fungi. In general nematodes have chitin substances in their egg shell rather than in other body segments (Clarke *et al.*, 1967; McClure and Bird, 1976; Perry and Tret, 1986).

a. Nematicidal chemicals

A number of phenolic compounds are produced by plants to fend off nematode attacks, helping the host suppress nematode invasion (Table 1). Chlorogenic acid produced in a wide range of solanaceous plants, rice and carrots has been reported to have nematicidal activity against a wide range of plant pathogenic nematodes (Pegard *et al.*, 2005). Research findings suggest that the activity of chlorogenic acid varies against different nematodes. It tends to show a moderately resistant reaction against the root knot nematode, M. incognita and is highly resistant against the false root knot nematode, Nacobbus aberrans. The hydrolysis of chlorogenic acid yields quinic acid and caffeic acid. The instability and toxic nature of these compounds to plants lead to a lack of correlation in effectiveness against nematodes. The phenolic compound, phenylphenaleone anigorufone accumulates at the infection site of Radopholus similis in resistant banana cultivars. This compound has also been found to be effective against the wilt fungus, F.oxysporum f.sp., cubense (Luis *et al.*, 1995).

Table 1. Phenolic compounds secreted by hosts along with their mode of action against PPNs

Compounds	Target nematodes	Mode of action	Reference
Kaempferol	*R. similis*	Inhibits egg hatching	Wyss *et al.* (1992)
Kaempferol, Quercetin, and Myricetin	*M. incognita*	Repellents / nemastatic	Wyss *et al.* (1992)
Medicarpin	*P.penetrans*	Inhibits motility	Baldridge *et al.* (1998)
Patuletin, Patulitrin, Quercetin, and Rutin	*H. zeae*	Inhibits juveniles	Faizi *et al.* (2011)
Glycelloin	*H.glycines*	Feeding inhibitor	Huang and Barker (1991)

b. Antagonistic plants

Antagonistic plants are plants that have the capacity to inhibit the activity of PPN through the production of metabolites. The metabolomics of antagonistic plants which includes both volatile and non-volatile compounds has been proven to exhibit nematicidal action against various PPN. The secretions of marigold plant roots contain α-terthienyl compounds which act as oxidative stress chemicals that have the capability to penetrate the nematode hypodermis (Gommers and Bakker, 1988; Wang *et al.*, 2007; Faizi *et al.*, 2011). Asparagusic acid of asparagus has been reported to show nematicidal activity against H. glycines and G. rostrochinensis (Takasugi *et al.*, 1975).

Another important plant family that exhibits greater nematicidal potential is Brassicaceae. This group of plants produces isothiocyanates and indole glucosinolate compounds that have the potential to reduce the population of PPN. Isothiocyanate compounds have been proven to have nematicidal activity against Meloidogyne sp. and major cyst nematodes by inhibiting egg hatching (Brown *et al.*,1997; Yu *et al.*, 2005). Camalexin another type of glucosinolate also has the potential to inhibit PPN.

c. Interuption of nematode chemotaxis

The compounds produced by the host plants have the capability to interrupt the PPN orientation towards the host roots. Ethylene over producing host roots showed more resistance against PPN by inhibiting the invasion of juveniles to host roots. These compounds are generally produced after wounding and pathogen invasion (Booker and DeLong, 2015; Marhavi *et al.*, 2019; Dakshinamoorhy *et al.*, 2014). The invasion of PPN leads to the secretion of ethylene in host roots which has the potential to suppress the further invasion of PPN. This reduced attractiveness is due to a reduction in the secretion of PPN attractant or over synthesis of PPN reductant.

Host cell wall reinforcement against PPN

Since the PPN targets the cell wall to gain entry into the host, it is a prerequisite that the host plants reinforce the cell wall structure in order to resist PPN attacks. Accumulation of lignin is considered one of the basal defence mechanisms against PPN invasion (Ji *et al.*, 2015; Huang *et al.*, 2016). Researchers have found that the higher accumulation of lignin in host roots is effective against M. incognita.

Chemical inducers such as β-aminobutyric acid (BABA), thiamine, and sclareol have been found to induce the accumulation of lignin in host plants. BABA is effective against Meloidogyne spp., by inhibiting the invasion of juveniles, delaying the giant cell formation and retarding the biology of nematodes. It has also been found that BABA helps to accumulate lignin in roots and callose in root galls induced by nematodes. Thiamine (Vitamin B) also triggers various defence pathways in plants such as activating the phenylpropanoid biosynthesis pathway, accumulating of lignin in roots and reducing the penetration and development of a wide range of PPN (Huang *et al.*, 2016). Treatment with sclareol also induces the accumulation of lignin in host roots. Sclareol has been found to be effective against a wide range of pathogens including fungi and bacteria (Fujimoto *et al.*, 2015; Bailey *et al.*, 1975; Kennedy *et al.*, 1992; Seo *et al.*, 2012)].

Callose deposition and suberin accumulation also helps to reinforce the host cell wall against various PPN. It has been found that upon infection by M. nassi, callose deposition is initiated in the early stages of infection and suberin accumulation is higher in the later stages of PPN infection in Aegilops variablis (Balhadere and Evans ,1995).

Lignin, suberin in suberin lamellae and casparian strips at the endodermis are found to be the basal level of defence against PPN especially, Meloidogyne sp., since these nematodes cannot have the capcity to cross the endodermis due to the reinforcement of the cell wall in casparian strips and suberin lamellae (Abad *et al.*, 2009; Wyss *et al.*, 1992).

Reactive Oxygen Species (ROS), Nitric oxide (NO), Protease Inhibitors Mediated Resistance Against PPN

Superoxide anion and hydrogen peroxide are the reactive oxygen species (ROS) produced in plants, which serve as early signalling molecules. These ROS are produced in plants during the early stages of PPN invasionand serve as signalling molecules to activate the immune response in plants such as cell wall strengthening and HR cell death (Torres *et al.*, 2006; Kadoda *et al.*, 2015). Tomato plants with the R gene (Mi1.2) show prolonged induction of ROS against M. incognita (Zhou *et al.*, 2018). Histochemical studies reveal that hydrogen peroxide accumulation is higher in nematode-infested apoplast (Melillo *et al.*, 2006).

The apoplastic ROS is produced by plasma membrane bound NADPH oxidase respiratory burst oxidase homologs (RBOH) (Kadoda *et al.*, 2015). The Mi1.2 mediated ROS accumulation is higher in plants with whitefly-induced 1 (WFI-1) which is a homolog of RBOH. This WFI-1 is regulated by HSFA1, a heat shock factor that binds to the WFI1 promoter region (Zhou *et al.*, 2018).

Nitric oxide (NO) serves as an important signalling molecule in plants. It has been found that tomato plants infected with M. incognita carrying the Mi1.2 gene accumulate more NO than susceptible plants (Melilo *et al.*, 2011). The application of sodium nitroprusside (SNP) is capable of inducing the NO synthesis in susceptible plants against M. incognita by reducing the fecundity of females, as SNP acts as an NO donor (Zhou *et al.*, 2015). NO is also involved in triggering Jasmonic acid-dependent pathways against PPN. JA and SNP have been found to silence protease inhibitor 2 which helps induce the immune response against PPN (Hepher and Atkinson, 1992). The accumulation of various protease inhibitors such as trypsin inhibitors and cysteine protease inhibitors confers resistance against PPN (Urwin *et al.*, 2000; Urwin *et al.*, 2003).

HR cell death against PPN

Hypersenstive reaction (HR) is a type of programmed cell death triggered after the invasion of PPN in host roots which helps the plants to stop the further spread of PPN (Huysmans *et al.*, 2017). The HR response is unique to pathogens such as fungi, bacteria, viruses and nematodes. In the case of PPN, HR death is observed in three phases of infection: i) at the cortex and epidermis during PPN penetration and migration, ii) at vascular tissues during feeding cell formation, and iii) at cells adjacent to feeding cells (Albuquerque *et al.*, 2010; Davis *et al.*, 2015a; Melillo *et al.*, 2006; Kim *et al.*, 2012). Most commonly HR cell death is induced during the formation of feeding sites of PPN. Tomato plants with Mi1.2 genes induce HR death during the formation of giant cells by inhibiting the feeding of nematodes (Melillo *et al.*, 2006).

HR cell death creates a gap between the feeding cells and surrounding cells which blocks the supply of water and nutrients. This mechanism is observed in tomato plants carrying the resistant gene Hero A which confers resistance against G. rostrochinensis. The resistant plants allow the formation of syncytia but the surrounding cells become dead resulting in the blockage of nutrients and water to the feeding site of nematodes (Sovczak *et al.*, 2005). Depending on the expression of host R genes, the mechanism of HR death varies. For instsnce, there may be direct death of the feeding site of cyst nematodes in soybean and root-knot nematodes in cowpea without typical HR cell death (Das *et al.*, 2008; Yan and Vaidoo, 2018).

Conclusion

It is a prerequisite to understand the response of plants corresponding to nematode invasion. The molecular aspects such as classes of receptors for receiving nematode-secreted proteins are still in the infancy stage. The real challenge in the near future in the field of nematology is the identification of root cause for triggering the immune response such as identification of proteins and their corresponding receptors and the lack of clarification in the field of signalling pathways related to host nematode interaction

References

Abad, P., Castagnone-Sereno, P., Rosso, M. N., Engler, J.D.A., and Favery, B. (2009). Invasion, feeding and development, In Root-knot nematodes. Eds. R. N. Perry, M. Moens, and J. L. Starr, Cambridge: CABI International Press, 163–181.

Albuquerque, E., Carneiro, R., Costa, P., Gomes, A., Santos, M., Pereira, A., et al. (2010). Resistance to Meloidogyne incognita expresses a hypersensitive-like response in Coffea arabica. Eur. J. Plant Pathol. 127(3), 365–373.

Bailey, J., Carter, G., Burden, R., and Wain, R. (1975). Control of rust diseases by diterpenes from Nicotiana glutinosa. Nature 255(5506), 328–329.

Baldridge, G. D., O'Neill, N. R., and Samac, D. A. (1998). Alfalfa (Medicago sativa L.) resistance to the root-lesion nematode, Pratylenchus penetrans: defense-response gene mRNA and isoflavonoid phytoalexin levels in roots. Plant Mol. Biol. 38(6), 999–1010.

Balhadère, P., and Evans, A. (1995). Cytochemical investigation of resistance to root-knot nematode Meloidogyne naasi in cereals and grasses using cryosections of roots. Fund. Appl. Nematol. 18(6), 539–547.

Booker, M. A., and DeLong, A. (2015). Producing the ethylene signal: regulation and diversification of ethylene biosynthetic enzymes. Plant Physiol. 169(1), 42–50.

Boutrot, F., and Zipfel, C. (2017). Function, discovery, and exploitation of plant pattern recognition receptors for broad-spectrum disease resistance. Annu. Rev. Phytopathol. 55, 257–286.

Brown, P. D., Morra, M. J., and Sparks, D. L. (1997). Control of soil-borne plant pests using glucosinolate-containing plants. Adv. Agron. 61, 167–231.

Chen, J., Hu, L., Sun, L., Lin, B., Huang, K., Zhuo, K., et al. (2018). A novel Meloidogyne graminicola effector, MgMO237, interacts with multiple host defence-related proteins to manipulate plant basal immunity and promote parasitism. Mol. Plant Pathol. 19(8), 1942–1955.

Chen, S., Chronis, D., & Wang, X. (2013). The novel GrCEP12 peptide from the plant-parasitic nematode Globodera rostochiensis suppresses flg22-mediated PTI. Plant Signaling & Behavior, 8(9), e25359.

Clarke, A. J., Cox, P. M., and Shepherd, A. M. (1967). The chemical composition of the egg shells of the potato cyst-nematode, Heterodera rostochiensis Woll. Biochem. J. 104(3), 1056–1060.

D'Ovidio, R., Mattei, B., Roberti, S., and Bellincampi, D. (2004). Polygalacturonases, polygalacturonase-inhibiting proteins and pectic oligomers in plant-pathogen interactions. Biochim. Biophys. Acta, 1696(2), 237–244.

Das, S., DeMason, D., Ehlers, J., Close, T., and Roberts, P. (2008). Histological characterization of root-knot nematode resistance in cowpea and its relation to reactive oxygen species modulation. J. Exp. Bot., 59(6), 1305–1313.

Davies, L. J., Brown, C. R., and Elling, A. A. (2015). Calcium is involved in the RMc1(blb)-mediated hypersensitive response against Meloidogyne chitwoodi in potato. Plant Cell Rep. 34(1), 167–177.

Dhakshinamoorthy, S., Mariama, K., Elsen, A., and De Waele, D. (2014). Phenols and lignin are involved in the defence response of banana (Musa) plants to Radopholus similis infection. Nematology 16, 565–576.

Faizi, S., Fayyaz, S., Bano, S., Iqbal, E. Y., Lubna, Siddiqi, H., et al. (2011). Isolation of nematicidal compounds from Tagetes patula L. yellow flowers: structure–activity relationship studies against cyst nematode Heterodera zeae infective stage larvae. J. Agric. Food Chem. 59(17), 9080–9093.

Fujimoto, T., Mizukubo, T., Abe, H., and Seo, S. (2015). Sclareol induces plant resistance to root-knot nematode partially through ethylene-dependent enhancement of lignin accumulation. Mol. Plant Microbe Interact. 28(4), 398– 407.

Gommers, F. J., and Bakker, J. (1988). Mode of action of alpha-terthienyl and related compounds may explain the suppressant effects of Tagetes species on populations of free living endoparasitic plant nematodes. Bioact. Mol. 7, 61–69.

Hepher, A., and Atkinson, H. J. (1992). Nematode control with proteinase inhibitors. EP 0502730 B1.

Hou, S., Liu, Z., Shen, H., and Wu, D. (2019). Damage-associated molecular pattern-triggered immunity in plants. Front. Plant Sci. 10, 646.

Huang, J. S., and Barker, K. R. (1991). Glyceollin I in soybean-cyst nematode interactions: spatial and temporal distribution in roots of resistant and susceptible soybeans. Plant Physiol. 96(4), 1302–1307.

Huang, W. K., Ji, H. L., Gheysen, G., and Kyndt, T. (2016). Thiamine-induced priming against root-knot nematode infection in rice involves lignification and hydrogen peroxide generation. Mol. Plant Pathol. 17(4), 614–624.

Huysmans, M., Lema A, S., Coll, N. S., and Nowack, M. K. (2017). Dying two deaths—programmed cell death regulation in development and disease. Curr. Opin. Plant Biol. 35, 37–44

Jaouannet, M., Magliano, M., Arguel, M. J., Gourgues, M., Evangelisti, E., Abad, P., et al. (2013). The root-knot nematode calreticulin Mi-CRT is a key effector in plant defense suppression. Mol. Plant Microbe Interact. 26(1), 97–105.

Ji, H., Kyndt, T., He, W., Vanholme, B., and Gheysen, G. (2015). β-aminobutyric acid-induced resistance against root-knot nematodes in rice is based on increased basal defense. Mol. Plant Microbe Interact. 28(5), 519–533.

Jones, J. D., and Dangl, J. L. (2006). The plant immune system. Nature 444 (7117), 323–329

Jones, J. T., Haegeman, A., Danchin, E. G., Gaur, H. S., Helder, J., Jones, M. G., et al. (2013). Top 10 plant-parasitic nematodes in molecular plant pathology. Mol. Plant Pathol. 14 (9), 946–961.

Kadota, Y., Shirasu, K., and Zipfel, C. (2015). Regulation of the NADPH oxidase RBOHD during plant immunity. Plant Cell Physiol. 56(8), 1472–1480.

Kennedy, B. S., Nielsen, M. T., Severson, R. F., Sisson, V. A., Stephenson, M. K., and Jackson, D. M. (1992). Leaf surface chemicals from Nicotiana affecting germination of Peronospora tabacina (adam) sporangia. J. Chem. Ecol. 18(9), 1467–1479.

Kim, Y., Kim, K., and Riggs, R. (2012). Initial subcellular responses of susceptible and resistant soybeans infected with the soybean cyst nematode. Plant Pathol. J. 28(4), 401–408.

Kud, J., Wang, W., Gross, R., Fan, Y., Huang, L., Yuan, Y., et al. (2019). The potato cyst nematode effector RHA1B is a ubiquitin ligase and uses two distinct mechanisms to suppress plant immune signaling. PLoS Pathog. 15(4), e1007720.

Lin, B., Zhuo, K., Chen, S., Hu, L., Sun, L., Wang, X., et al. (2016). A novel nematode effector suppresses plant immunity by activating host reactive oxygen species-scavenging system. New Phytol. 209(3), 1159–1173.

Lozano-Torres, J. L., Wilbers, R. H., Gawronski, P., Boshoven, J. C., Finkers- Tomczak, A., Cordewener, J. H., et al. (2012). Dual disease resistance mediated by the immune receptor Cf-2 in tomato requires a common virulence target of a fungus and a nematode. Proc. Natl. Acad. Sci. U S A 109(25), 10119–10124.

Luis, J. G., Fletcher, W. Q., Echeverri, F., Abad, T., Kishi, M. P., and Perales, A. (1995). New phenalenone-type phytoalexins from Musa acuminata (Colla AAA) grand nain. Nat. Prod. Lett. 6(1), 23–30.

Manosalva, P., Manohar, M., Von Reuss, S. H., Chen, S., Koch, A., Kaplan, F., ... & Klessig, D. F. (2015). Conserved nematode signalling molecules elicit plant defenses and pathogen resistance. Nature communications, 6(1), 1-8.

Marhavý, P., Kurenda, A., Siddique, S., Denervaud Tendon, V., Zhou, F., Holbein, J., et al. (2019). Single-cell damage elicits regional, nematode-restricting ethylene responses in roots. EMBO J. 38(10), e100972.

McClure, M. A., and Bird, A. F. (1976). The tylenchid (Nematoda) egg-shell: formation of the egg-shell in Meloidogyne javanica. Parasitology 72(1), 29–39.

Melillo, M. T., Leonetti, P., Bongiovanni, M., Castagnone-Sereno, P., and Bleve-Zacheo, T. (2006). Modulation of reactive oxygen species activities and H2O2 accumulation during compatible and incompatible tomato-root-knot nematode interactions. New Phytol. 170(3), 501–512.

Melillo, M., Leonetti, P., Leone, A., Veronico, P., and Bleve-Zacheo, T. (2011). ROS and NO production in compatible and incompatible tomato-Meloidogyne incognita interactions. Eur. J. Plant Pathol. 130(4), 489–502.

Mendy, B., Wang'ombe, M. W., Radakovic, Z. S., Holbein, J., Ilyas, M., Chopra, D., et al. (2017). Arabidopsis leucine-rich repeat receptor–like kinase NILR1 is required for induction of innate immunity to parasitic nematodes. PLoS pathogens, 13(4), e1006284

Naalden, D., Haegeman, A., de Almeida-Engler, J., Birhane Eshetu, F., Bauters, L., and Gheysen, G. (2018). The Meloidogyne graminicola effector Mg16820 is secreted in the apoplast and cytoplasm to suppress plant host defense responses. Mol. Plant Pathol. 19(11), 2416–2430.

Pegard, A., Brizzard, G., Fazari, A., Soucaze, O., Abad, P., and Djian-Caporalino, C. (2005). Histological characterization of resistance to different root-knot nematode species related to phenolics accumulation in Capsicum annuum. Phytopathology 95(2), 158–165.

Perry, R. N., and Trett, M. W. (1986). Ultrastructure of the eggshell of Heterodera schachtii and H. glycines (Nematoda: Tylenchida). Rev. Nématol. 9(4), 399–403.

Seo, S., Gomi, K., Kaku, H., Abe, H., Seto, H., Nakatsu, S., et al. (2012). Identification of natural diterpenes that inhibit bacterial wilt disease in tobacco, tomato and Arabidopsis. Plant Cell Physiol. 53 (8), 1432–1444.

Sobczak, M., Avrova, A., Jupowicz, J., Phillips, M. S., Ernst, K., and Kumar, A. (2005). Characterization of susceptibility and resistance responses to potato cyst nematode (Globodera spp.) infection of tomato lines in the absence and presence of the broad-spectrum nematode resistance Hero gene. Mol. Plant Microbe Interact. 18(2), 158–168.

Takasugi, M., Yachida, Y., Anetai, M., Masamune, T., and Kegasawa, K. (1975). Identification of asparagusic acid as a nematicide occuring naturally in the roots of asparagus. Chem. Lett. 1, 43–44.

Torres, M. A., Jones, J. D., and Dangl, J. L. (2006). Reactive oxygen species signaling in response to pathogens. Plant Physiol. 141(2), 373–378.

Urwin, P. E., Green, J., and Atkinson, H. J. (2003). Expression of a plant cystatin confers partial resistance to Globodera, full resistance is achieved by pyramiding a cystatin with natural resistance. Mol. Breed 12(3), 263–269.

Urwin, P. E., Levesley, A., Michael J. McPherson, M. J., and Atkinson, H. J. (2000). Transgenic resistance to the nematode Rotylenchulus reniformis conferred by Arabidopsis thaliana plants expressing proteinase inhibitors. Mol. Breed 6(3), 257–264.

Wang, K.H., Hooks, C. R., and Ploeg, A. (2007). Protecting crops from nematode pests: using marigold as an alternative to chemical nematicides, In Plant Disease; PD-35 (Honolulu: University of Hawaii).

Wyss, U., Grundler, F. M. W., and Munch, A. (1992). The parasitic behaviour of second-stage juveniles of Meloidogyne incognita in roots of Arabidopsis thaliana. Nematologica 38, 98–111.

Wyss, U., Grundler, F. M. W., and Munch, A. (1992). The parasitic behaviour of second-stage juveniles of Meloidogyne incognita in roots of Arabidopsis thaliana. Nematologica 38, 98–111.

Yan, G. P., and Baidoo, R. (2018). Current research status of Heterodera glycines resistance and its implication on soybean breeding. Engineering 4(4), 534–541.

Yang, S., Dai, Y., Chen, Y., Yang, J., Yang, D., Liu, Q., et al. (2019). A novel G16B09- like effector from Heterodera avenae suppresses plant defenses and promotes parasitism. Front. Plant Sci. 10, 66.

Yu, Q., Tsao, R., Chiba, M., and Potter, J. (2005). Selective nematicidal activity of allyl isothiocyanate. J. Food Agr. Environ. 3(2), 218–221.

Zhou, J., Jia, F., Shao, S., Zhang, H., Li, G., Xia, X., et al. (2015). Involvement of nitric oxide in the jasmonate-dependent basal defense against root-knot nematode in tomato plants. Front. Plant Sci. 6, 193.

Zhou, J., Xu, X. C., Cao, J. J., Yin, L. L., Xia, X. J., Shi, K., et al. (2018). Heat shock factor HsfA1a is essential for R gene-mediated nematode resistance and triggers H2O2 production. Plant Physiol. 176(3), 2456–2471.

11

Persistence of Herbicides in Crops and Its Management

Dhamodharan P.[1], K. Kiruthika[2], M. Raghuramakrishnan[1] G. Gunalarasi[3] and S. Somasundaram[4]

[1]*Department of Agronomy, Tamil Nadu Agricultural University Coimbatore, Tamil Nadu*

[2]*Department of Agronomy, Vanavarayar Institute of Agriculture Pollachi, Tamil Nadu*

[3]*Department of Agronomy, Dhanalakshmi Srinivasan University Trichy*

[4]*Cotton Research Station, Veppanthattai, Tamil Nadu*

Abstract

Herbicides play a pivotal role in modern agriculture, aiding in weed control and ensuring optimal crop yield. However, the persistence of these chemicals poses challenges, impacting subsequent crop cycles and raising environmental concerns. Analysing the factors influencing herbicide persistence, including soil type, climate, and chemical properties, reveals the complex interplay that contributes to extended residual activity. Furthermore, this chapter provides insights into innovative management approaches aimed at mitigating herbicide persistence issues. These may include alternative herbicide formulations, application techniques, and sustainable agricultural practices that minimise the ecological footprint. Additionally, advancements in precision agriculture and integrated pest management are discussed as promising avenues for optimising herbicide application, reducing residues, and promoting sustainable farming practices. By shedding light on the persistence of herbicides in crops and proposing effective management strategies, this study contributes to the ongoing dialogue on sustainable agriculture, offering insights that can guide farmers, policymakers, and researchers towards environmentally responsible and economically viable practices.

Keywords: *Ecological footprint, Herbicide persistence, Weed control, Sustainable agriculture*

Introduction

Agrochemicals are categorised based on the specific target organisms they are intended to control, including insects, weeds, and fungi. Among these target organisms, weeds significantly contribute to the highest economic losses due to their interference in crop production. Herbicides, the predominant class of agrochemicals globally, account for 48% of the total expenditure. The largest consumers of agrochemicals are Europe, Asia, and the United States. Herbicides have various biological and chemical properties (Roy *et al.*, 2023a). They exhibit variations in behaviour and spectrum weed control. These properties play an important role in herbicide selection, method of utilisation and application, post-application effects such as toxicity and residue accumulation in the soil. The crucial factor for any herbicide application is the behaviour of the herbicide in the soil and its transformation or modification once it reaches the soil (Miller and Westra, 1914). The fate of herbicides and their efficiency are regulated by herbicide properties, soil characteristics and climatic conditions.

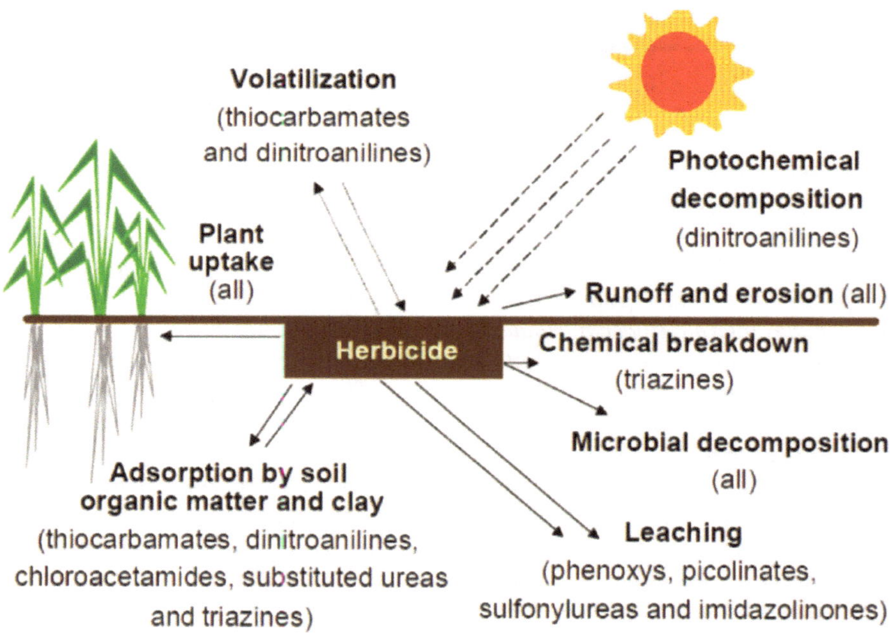

Fig. 1: Fate of herbicides in environment and processes affecting their action

Fate of Herbicide in Soil

- Processes affecting the fate of herbicides

There are mainly three processes which affect the fate of herbicides. These are transport and degradation processes and chemico-physico properties of herbicides which are detailed below:

1. Volatalisation: Volatilisation is the process by which a substance changes from a solid or liquid state to a gaseous state. All chemicals have a finite vapour pressure. Volatilisation is similar to the evapouration and vapourisation process. Vapour pressure is the pressure exerted by the gas particles of an herbicide when in equilibrium with the liquid or solid state in a closed container.

Vapour pressure influences volatility, with higher vapour pressure, indicating greater volatility. The general rule is that if the vapour pressure of the herbicide is >1x10-4 (0.0001) mm of Hg at 25°°C it will need to be incorporated to prevent loss due to volatility or evapouration. For example, EPTC is approximately 300 times more volatile than trifluralin. Incorporation of trifluralin is not due to volatilisation, but rather to photodegradation (breakdown by sunlight).

Table 1. Examples of some herbicides with vapour pressure in mm of Hg at 25 °C

Herbicide	**Vapour pressure in mm of Hg @ of 25 °C**
EPTC, Eptam, Eradicane	3.4 x 10-2 (0.034)
Vernolate, Verman	1 x 10-2 (0.01)
Clomazone, Command	1.4 x 10-4 (0.00014)
Trifluralin, Treflan/Tri-4	1.1 x 10^{-4} (0.00011)
Atrazine	2.9 x 10-7 (0.00000029)
Paraquat, Gramaxone	TLTL (Too low to list)

Volatilisation can be both negative and positive:

- Negative: Herbicide drift, leading to inadequate weed control.
- Positive: The herbicide penetrates the soil, reaching depths of up to 4 inches for EPTC and approximately 1 inch for trifluralin. This movement within the soil fills pore spaces, facilitating even distribution and promoting uniformity. Consequently, this enhances the efficacy of weed control.

Factors Affecting Volatilisation

- Temperature: Increased temperatures result in higher volatility. The least favorable time to apply a volatile herbicide, considering loss due to volatility, is during midday. However, it is anticipated that weed control effectiveness would likely increase during this period.

- Soil moisture: High soil water content increases volatility (water vapour acts as a carrier)

Methods to Decrease Volatilisation

- Soil incorporation (traps herbicide within the soil and prevents loss)
- Use of special formulations - granular, micro-encapsulated, salt (2,4-D); ester formulations most volatile, salt formulation least volatile
- Apply when conditions are unfavorable for volatilisation (e.g., lower temperature)

2. **Photodecomposition:** UV light's impact and the process of photodecomposition involve the breakdown of herbicides under sunlight, specifically ultraviolet light (UV peaks at 400 nm). When a herbicide absorbs light in the UV range, its molecular structure undergoes change, resulting in the breakage of chemical bonds. While all herbicides are susceptible to photodecomposition, the extent of this effect varies. In their molecular form, most herbicides appear white or nearly white, with peak light absorption occurring in the 220-324 nm range. In contrast, dinitroaniline herbicides (DNA) present as yellow in their molecular form (referred to as "yellow herbicides"), and their peak light absorption is approximately 376 nm, much closer to the peak of UV light (400 nm). This elucidates why yellow herbicides are more prone to photodegradation.

Almost all herbicides are susceptible to photodecomposition including dinitroanilines (DNAs), S- triazines, and ureas. The presence of riboflavin in aqueous solution resulted in a decrease in photodegradation, leading to an increase in Imazaquin half-life compared to pure water. The presence of humic acids delayed the photodegradation of Imazethapyr, but titanium dioxide increased the rate of photodegradation.

Methods to Decrease (Photodecomposition)

- Movement into soil - incorporation, rainfall, diffusion - a soil applied herbicide must get into the soil solution to be active
- Surface movement in water - runoff of herbicide can be detrimental to the environment (non-point source pollution). Adsorbed (bound/tied up) to organic matter (fate on soil surface).
- Electrical attraction- organic matter has a strong negative (-ve) charge and attracts cations which are positive (+ve). The charge status of an herbicide may be positive, negative, or neutral.

- Adsorptive Capacity (CEC)- This denotes the soil's ability to attract cations. A higher CEC signifies an increased number of binding sites available for the herbicide. Elevated levels of organic matter on the soil surface may lead to herbicide binding, resulting in inadequate weed control. This phenomenon is commonly observed with certain herbicides.

3. **Leaching:** Downward movement of herbicide into a soil profile, movement can be both lateral and reverse leaching (upward movement) occur when a wet soil dries. Leaching is essential for soil-applied herbicides to transport into the zone where weed seeds are located. Leaching can result in herbicide movement out of the zone where weed seed will germinate (loss of herbicide activity) and may increase crop injury if moved too deep (Sow and Ranjan, 2020).

Factors Influencing Leaching

- Soil texture, permeability, rainfall, adsorption
- Soil permeability would be affected by the presence or absence of a hardpan that could restrict leaching
- The greater the rainfall the greater the potential for leaching. Herbicides adsorbed to soil or organic matter would not be susceptible to leaching.
- The higher the water solubility of a herbicide, the increased probability of leaching. This tendency is also influenced by the adsorptive capacity of the herbicide.

4. **Adsorption/desorption to soil particles/organic matter:** Adsorption is the association of a herbicide with the surface of a soil particle. An equilibrium exists between the soil particle and the soil solution and adsorption and desorption are both occurring. Due to electrical attraction, a positively charged herbicide (+ve) binds to negatively charged clay particles (-ve), resulting in immobilisation. Adsorbed herbicide, in this state, is unavailable for weed control; the herbicide must be in the soil solution to exert its weed control properties. The adsorptive capacity is herbicide-dependent, and the soil organic carbon sorption coefficient (KOC value) serves as a comparative measure for herbicides. A higher adsorptive capacity indicates a lower potential for leaching. Herbicide is bound to soil or organic matter and not in the soil solution where it can be leached. As the solubility of the herbicide decreases, sorption increases. Desorptive capacity (CEC) in clay, sand and organic matter is the ability of the soil to attract cations (Takeshita *et al.*, 2019). It means the higher the CEC, the more binding sites available. Clay soils

have a higher CEC than sandy loam soil. A soil with a high organic matter content [a lot of negative sites (-ve)] will also bind herbicides and prevent them from being present in the soil solution.

a) **Effect of soil texture/organic matter on herbicide rate** - For some herbicides, a higher herbicide rate is required for effective weed control in clay soils compared to sandy loam soils. In clay soil, the herbicide concentration needs to be sufficiently high to "saturate" all binding sites while ensuring an ample amount remains in the soil solution for effective weed control. (Zia-Ul-Haq *et al.*, 2019).

b) **pH effect on certain herbicides and adsorption** - Soil pH can alter the ionic characteristics of a herbicide and impact adsorption. In low pH (acidic) soil, Atrazine becomes cationic (+charge), leading to its adsorption onto soil particles and rendering it unavailable for weed control. As a result, Atrazine exhibits increased activity in soils with higher pH levels (between pH 6 to 8 compared to pH 5). In these conditions, it remains uncharged, making it available in the soil solution for root uptake.

5. **Hydrolysis:** Hydrolysis is the reaction of the herbicide with water (H_2O). A hydrogen ion (H+) or more commonly a hydroxyl group (-OH) is added to the molecule. The structure of the herbicide is changed chemically. It does not take much change in a chemical structure to significantly affect activity. For example, in simazine when chlorine is replaced with an OH group to form hydroxysimazine, toxicity decreased by 1000 times.
6. **Microbial decomposition of herbicides:** Microbes have enzymes that are secreted and act as catalysts to drive specific reactions (microbial / biochemical)

- Bacteria - procaryotes single cell microorganisms, with no organised nucleus
- Actinomycetes - bacteria that form branching filaments with a higher level of development compared to bacteria.
- Fungi - eukaryotes with a higher development level than bacteria or actinomycetes

Mechanism of Herbicide Degradation

- **Dehalogenation** - removal of halogen from the herbicide molecule. Halogens are those elements toward the right side of the Periodic Table, Fluorine (F), Chlorine (Cl), Bromine (Br), and Iodine (I)

- **Dealkylation** - removal of an alkyl group (C-H aliphatic chain).
- **Amide hydrolysis** - cleavage of a bond of an amide (contains NH2 group) by addition of water, a weak acid and NH3 is produced.
- **Ester hydrolysis** - cleavage of a bond of an ester by addition of water, a weak acid is produced.
- **Beta-oxidation** - a normal pathway for degradation of long chain fatty acids, 2 carbons are removed when a compound goes through the pathway once.
- **Ring hydroxylation** - reaction with water whereby a OH group is added to the ring structure.
- **Ring cleavage** - the bonds within the ring structure of the herbicide molecule are split or cleaved.
- **Reduction of nitro groups in anaerobic conditions** - NO2 is reduced to NH2 and O2 is released.
- **Moisture** - 50% to 100% of field capacity is optimum.
- **Temperature** - 80 – 90°F ideal.
- **pH** - bacteria/actinomycetes predominate at pH >5.5 fungi predominate at pH < 5.5 competition can occur among microorganisms.

Herbicide Persistence

The persistence of a herbicide is defined as the time during which the molecule remains in the soil and is usually expressed as half-life. Half-life (t1/2) refers to the time required for half of the herbicide to decompose and disappear from the soil. The length of time a herbicide remains active in the soil is called soil residual life or soil persistence. On agricultural land a major problem arising from herbicide persistence is injury to rotation crops. The half-life of a herbicide in the field will vary depending on environmental and soil conditions. As an illustration, a herbicide with a half-life of 4 weeks would undergo 50 percent decomposition, leaving 50 percent of it present in the soil four weeks after application. After 8 weeks, 75 percent of the herbicide would be degraded, with 25 percent remaining, and after 12 weeks, only 12.5 percent would remain.

Notorious herbicides such as triazine herbicides can injure rotation crops even 6 to 24 months after their application at normal rates. Several substituted ureas, sulphonylureas, dichlobenil, diphenamid, and 2,3,6 – TBA also often pose phytotoxic residue problems on crop land. Even a short residue herbicide like glyphosate has been reported to damage tomato transplants.

Factors Affecting Decomposition of Herbicides are

- Volatalisation
- Decomposition
- Plant uptake
- Surface runoff
- Leaching
- Soil Absorption
- Chemical decomposition
- Microbial decomposition

Table 2. Half-life in soil and water for common herbicides

Active Ingredient	Product	½ Life in soil	½ life in water
2,4-D	Surpass (PB)	< 7 days	1 – several weeks
2,4 -DB	Buttress	7 days	
MCPA	MCPA 500	14 – 31 days (3 – 4 months residual)	2 – 5 weeks
Aciflourfen	Blazer	59 days	4 days
Atrazine	Gesasprim (PB) (W)	60 - > 100 days (up to 1 year if dry)	55 days (range 50 – 105 days)
Cyanazine	Bladex	2 – 14 weeks	
Dicamba	Banvel	1– 4 weeks	
Diquat	Reglone	> 1000 days (remain in top 25 mm)	< 2 days
Diuron	Diuron	90 days (range 1 month – 1 year)	
Glyfosinate	Basta	7 – 20 days	2 – 30 days
Glyphosate	Roundup dry (w)	47 days (range 3 – 130 days) no pre-emergent activity	35 – 63 days
Imazapyr	Onduty (PB)	6 months – 2 years residual (temperate); 3 – 6 months (crop)	
Metalochlor	Clincher (PB)	114 days (range 15 – 132 days)	97 - > 200 days (alkaline – acid)
Metribuzin	Lexine (PB)	60 days (range 30 – 120 days)	7 days
Metsulfuron	Ally (PB)(W)	30 days (range 14 – 180 days)	29 - > 84 days (increasing in concentration)
Oxyflourfen	Goal (W)	30 – 40 days	Adsorbed by sediments

Paraquat	Gramaxone (W)	> 1000 days (adsorbed by soil and organic matter – not available to plants)	30 days – 23 weeks (mostly adsorbed by sediments)
Pendimethalin	Stomp (PB) (W)	40 days (persistent for 3 – 4 months)	
Picloram	Tordon granules (PB)	90 days (range 20 – 300 days increasing in application rate)	2.6 days
Simazine	Gesatop (PB) (W)	60 days (range 28 – 149 days; 1 year resildual in high pH soil)	30 days
Sulfosulfuron	Monza (PB)	24 days (range 11- 47 days)	
Trifluralin	Treflan (PB)	57 – 126 days (6- 8 months resildual)	

Herbicide Residue Management

There are two main approaches in herbicide residue management

1. Indirect Approach

- The use of chemical activators and surfactants
- Selection of the most appropriate time and method of application of the herbicide
- Selective stimulation of crops to smother the weeds and
- Supplementing herbicidal control with tillage and other methods of weed control (Roy *et al.*, 2023b)

2. Direct Approach

a. Molecular structure approach

- At equal rates, volatile, photo decomposable and biodegradable herbicide structures are less persistence than the other types of structures.
- In the case of volatile herbicides, their dilution in air is sought as solution to their long persistence in soil.
- Thus developing volatile, photo decomposable and biodegradable structures of soil active herbicides should prove helpful in lowering at least the soil pollution.

b. Controlled mobility approach

- Mobility of herbicides in soil can be manipulated to dissipate the herbicide by leaching and volatilisation
- Heavy irrigation – mobile herbicide beyond the root zone- contribute to Ground water

- Repeated deep tillage of the treated soil will hasten both volatile loss and degradation of herbicides and it may also dilute the herbicide by mixing it with a larger volume of soil to innocuous levels (Dayal *et al.*, 2023).

c. Enhanced degradation approach

- Herbicide triazines can be degraded faster when the soil pH is lowered by the application of sulphur, but this may not be permissible under all conditions.
- Sodium and calcium polysulphides have been found effective in deactivating simazine and atrazine in lab research by replacing chlorin radicals of the herbicide ions by adding 50 t/ha of muck and peat added to soil may often eliminate triazine residues from the soil.

d. Toxicity prevention approach

- Adding organic matter to sandy soil has been reported to reduce the herbicidal activity of fluometuron and fenuron in growth chamber studies, as well as in field experiments.
- Addition of P to soil reduce the phytotoxicity to diuron to cotton.

e. Trap - crop approach

- Herbicide residue problems in soils can be addressed by deliberately including crop plants in rotation that are resistant to a particular herbicide.
- This permits the land to remain in production and, at the same time, it hastens the dissipation of the herbicide such crops are called herbicide trap crops
- E.g. sugarcane, sorghum and maize are good trap crops for eliminating residues of atrazine and simazine from soil.

Future Prospects

In India, the primary focus of research has been on the cultural and mechanical management of herbicide residues in soil, with some exploration into split application or rotational use of herbicides. However, there is a notable scarcity or absence of research on herbicide deactivation using various organic sources, enhancing degradation through biostimulation, and employing phytoremediation techniques for contaminant removal from sites. Additionally, there is limited understanding of how crop residues influence the behavior of herbicide residues in the soil and the broader environment. To address these knowledge gaps, it is imperative to undertake extensive, site-specific field studies. These studies are crucial for developing comprehensive strategies to manage herbicide residues in the soil environment. By expanding the scope of research to encompass these critical aspects, we can advance our understanding and contribute to the development of holistic measures for effective herbicide residue management.

Conclusion

Herbicides play a vital role in crop production programs, yet their indiscriminate use can pose risks to the soil environment. To ensure sustainability and prevent environmental degradation, it is essential to consider all aspects of prevention and management while employing herbicides. Striking a balance between achieving abundant harvests and maintaining high-quality food production is crucial. Therefore, integrating mechanical and cultural management practices with herbicide use emerges as a practical option. To address issues related to herbicide persistence, the initial step is to opt for less persistent products. Before selecting any herbicide, carefully review the label for recrop statements. Ensuring the appropriate rates and application timings for all pesticides is imperative. Implementing selective tillage, utilising herbicide combinations, and choosing tolerant crops and varieties can further mitigate the risk of carryover crop injury. By adopting a judicious approach to herbicide use, we contribute to the continued availability of these essential weed management tools for the future. This strategy promotes sustainable agriculture, fostering both productive harvests and environmental responsibility.

References

Dayal, P., Kumar, A., Tyagi, S., Pal, R.K., Manohar, B., Ranjan, S. and Sow, S. (2023). Weed dynamics and productivity of wheat (Triticum aestivum) under various tillage and weed management practices. The Indian Journal of Agricultural Sciences 93(9), 1037–1040.

Miller, P. and Westra, P. (1914). Production, Herbicide behaviour in soils. Crop series. No. 0.562.

Roy, D. K., Singh, D., Sow, S. and Ranjan, S. (2023a). Impact of Weed Management Practices On Weed Dynamics and Productivity of Rice-Wheat Cropping System. Scientist doi:10.5281/zenodo.7573159

Roy, D.K., Sow, S. and Ranjan, S. (2023b). Productivity and profitability of zero till winter maize as influenced by integrated weed management practices. Indian Journal of Weed Science 55(3), 264-268.

Sow, S. and Ranjan, S. (2020). Brown Manuring, an Effective Technique for Weed Management and Sustainable Yield of Cereal Crops. Vigyan Varta 1(4), 32-34.

Takeshita, V., Mendes, K.F., Alonso, F.G. and Tornisielo, V.L. (2019). Effect of organic matter on the behaviour and control effectiveness of herbicides in soil. Plant Daninha 37, 11-17.

Zia-Ul-Haq, M., Khaliq, A., Sheng, Q., Matloob, A., Hussain, S., *et al.* (2019). Weed growth, herbicide efficacy, and rice productivity in dry seeded paddy field under different wheat stubble management methods. Journal of Integrative Agriculture 18(4), 907–926.

12

Precision and Data-Driven Farming for Sustainable Agriculture

Santosh Shinde, Pundlik Waghmare
Sharad Chenalwad and Anil Chavan

Department of Agronomy, Vasantrao Naik Marathwada Krishi Vidyapeeth Parbhani (MS), India

Abstract

The agricultural sector is undergoing a profound shift marked by digitalisation and automation, revolutionising farming practices through mechanisation, computerisation, and the pivotal role of precision agriculture. This transformation responds to a host of challenges: escalating production costs, the impact of climate change, and labor shortages among them. Covering about 38% of the Earth's surface, agriculture faces limitations, with substantial portions of land unsuitable for cultivation, including deserts, mountains, and urban areas. Between 1985 and 2005, agricultural land expanded by a mere 2.4%, primarily due to unsustainable practices like deforestation in tropical regions. This method cannot be a sustainable solution. The world's food demand is projected to double between 2005 and 2050. The Food and Agriculture Organisation of the United Nations (UN) estimates a need for 60% more food by 2050, largely driven by population growth. In 2012, the global population exceeded 7 billion, and though the growth rate has slowed, the UN predicts a population of 9.6 billion by 2050. Meeting this demand necessitates enhancing the productivity of existing fields. Increasing the yield of 16 key crop types to 95% of their potential could lead to a remarkable 58% increase in food production. Precision agriculture, powered by data-driven methodologies, plays a pivotal role in this pursuit by addressing field variations and optimising resources to achieve these goals.

Keywords: *Automation, Digitalisation, GPS, GIS, VRT, Mechanisation Precision agriculture, Sustainable agriculture*

Introduction

Agriculture faces a massive challenge: by 2050, it needs to sustainably feed over 9 billion people while supporting economic growth and protecting the environment. Meeting this demand requires a staggering 70% increase in food production from 2006 levels, mainly by making the most of existing farmland and resources.

Precision agriculture offers a promising way forward. It's about smartly managing soil and crops using technology and information to boost profits while being eco-friendly. This approach involves giving the precise amount of resources—like water, nutrients, and pesticides—at the right time and place to optimise farming. It's all about understanding and using the differences within fields to improve crop growth and protect the environment. Traditionally, understanding these differences involved collecting tons of soil and plant samples for lab analysis. But this can be costly. So, finding better, more efficient ways to assess these differences in soil and plant growth is crucial.

Precision agriculture relies on high-tech tools like GPS, GIS, yield monitors, and remote sensing to measure differences in soil and crop growth across fields. By integrating these technologies with tools that adjust resource usage based on these differences (called variable rate technologies), farmers can apply fertilisers and water precisely where they're needed, optimising crop yields (Ranjan *et al.*, 2022a). This method has three main parts: collecting data, understanding that data, and then making smart decisions based on it, all tailored to specific areas and timings. New technologies like yield monitors, satellites, and handheld sensors have made data collection faster and easier. But the real challenge is making sense of all this information and using it effectively for better farming. This review delves into how these technologies can enhance agricultural production and the hurdles they bring.

Precision and Data Driven Farming

Precision agriculture technology generates extensive farmland data that necessitates effective management, analysis, and processing. These data serve as the foundation for creating prescription maps, guiding crucial decision-making in farming practices. The generation of prescription maps and subsequent decision-making isn't solely about graphic processing and calculations; it also involves incorporating experience and knowledge into the process. Finding ways to effectively utilise information technology to support variable-rate operations stands as a pivotal task in precision agriculture.

Fig. 1: The revolution of precision and data-driven farming

Data sources in precision agriculture encompass both antecedent data and real-time data. Antecedent data comprises historical information such as multi-year yield data and soil nutrition data. Real-time data, on the other hand, encompasses live, current information obtained through various means:

1. Remote Sensing Data: Gathered from airborne and spaceborne sources, this data aids in the inversion of chlorophyll and Leaf Area Index (LAI), providing insights into crop health and growth.
2. Plant Growth Monitoring: Data reflecting the current status of plant growth, such as SPAD readings (chlorophyll content) and Leaf Area Index, contribute to real-time assessments of crop health and vigor.

Integrating these diverse data sources facilitates the creation of comprehensive maps and actionable insights. This enables farmers to make informed decisions about resource allocation, precisely applying inputs like fertilisers and water where and when they are most needed. The process involves not only the technical aspects of data processing and analysis but also the incorporation of farmers' experience and domain knowledge into the decision-making process. Ultimately, it's about leveraging technology to optimise farming practices, ensuring better yields, resource efficiency, and environmental sustainability in agriculture (Mohapatra *et al.*, 2023).

Precision Farming for Sustainable Agriculture

Precision farming, also known as precision agriculture, is a comprehensive approach that integrates technology, data analysis, and tailored management practices to optimise agricultural production while minimising environmental impact.

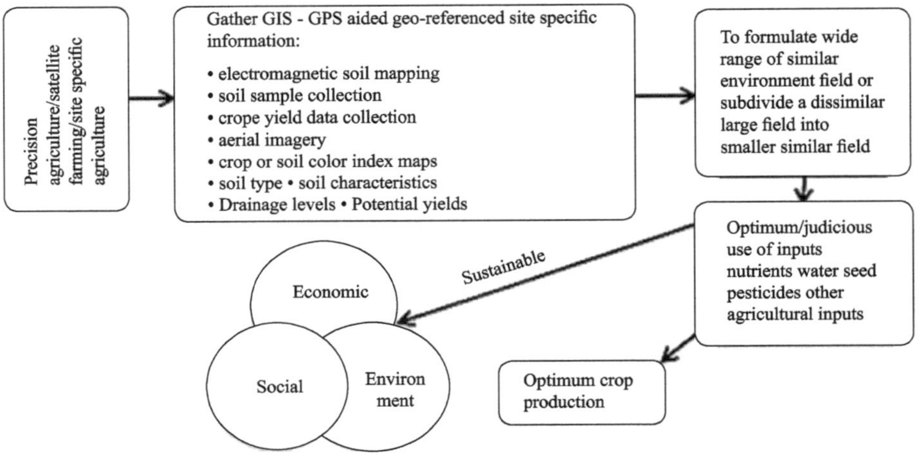

Fig. 2: Precision farming for sustainable agriculture

Its primary goal is to achieve sustainable farming by maximising yields, minimising input usage (such as water, fertilisers, pesticides), reducing waste, and preserving natural resources. In detailed as

1. Data Collection and Analysis

- Remote Sensing and GIS: Satellite imagery, drones, and Geographic Information Systems (GIS) are used to gather data on field variability, including soil properties, topography, and crop health.
- Sensor Technology: On-the-go sensors and devices collect real-time data on soil moisture, nutrient levels, and plant health.

2. Field Mapping and Analysis

Information collected is analysed to create detailed field maps that highlight variations in soil fertility, moisture levels, and crop health. This helps identify areas that need specific attention.

3. Precision Application of Inputs

- Variable Rate Technology (VRT): Utilises data-driven insights to adjust and apply resources like fertilisers, pesticides, and water precisely where and when they're needed. This prevents overuse in some areas and underuse in others, optimising crop growth and minimising environmental impact (Sow *et al.*, 2023).
- Auto-steer Systems: GPS-guided systems on farming equipment ensure precise and efficient application of inputs while minimising overlap.

4. Decision Support Systems

Software tools and models process the collected data to provide farmers with actionable insights. These systems help in making informed decisions about crop rotations, planting schedules, and resource allocation.

5. Yield Monitoring and Analysis

Yield monitors track and record crop yields during harvesting, allowing farmers to correlate yield variations with specific field conditions. This data aids in refining future management practices.

6. Environmental Conservation

Precision farming emphasises reducing environmental footprint by minimising the use of chemicals, conserving water resources, and preventing soil erosion through targeted application and management.

7. Economic Viability

By optimising input usage and maximising yields, precision farming aims to enhance profitability for farmers while ensuring sustainable agricultural practices.

8. Continuous Improvement

The data-driven nature of precision farming allows for continuous monitoring, analysis, and adaptation. Farmers can refine their practices based on ongoing insights, thereby improving efficiency and sustainability over time.

Precision farming stands as a key approach to sustainable agriculture, enabling farmers to produce more with fewer resources, reduce environmental impact, and foster long-term agricultural sustainability in the face of growing food demand and resource constraints.

Benefits of Precision Agriculture

1. Increased Productivity: By optimising resource use and tailoring treatments to specific areas of a field, precision agriculture can increase crop yields.
2. Resource Efficiency: Precision agriculture reduces the overuse of fertilisers, pesticides, and water, which is not only cost-effective but also environmentally friendly.
3. Cost Savings: Lower input costs and increased productivity result in cost savings for farmers.

4. Environmental Sustainability: Reduced resource wastage and more efficient farming practices contribute to sustainable agriculture and a reduced environmental impact.
5. Data-Driven Decision-Making: Farmers can make informed decisions based on real-time data, leading to better outcomes and crop management.

Overall, precision agriculture represents a significant shift in the way farming is conducted, leveraging technology and data to make agriculture more efficient, sustainable, and profitable.

Data Driven Farming for Sustainable Agriculture

Data, information, and knowledge play pivotal roles in agriculture, and precision agriculture specifically relies on the data produced within the farm itself. This distinguishes it from external data sources that may hold value for producers but are collected and disseminated by other entities. These concepts tie closely to the notion of data-driven farming, a burgeoning field that extends beyond technological advancements and has garnered significant attention and investment in precision agriculture.

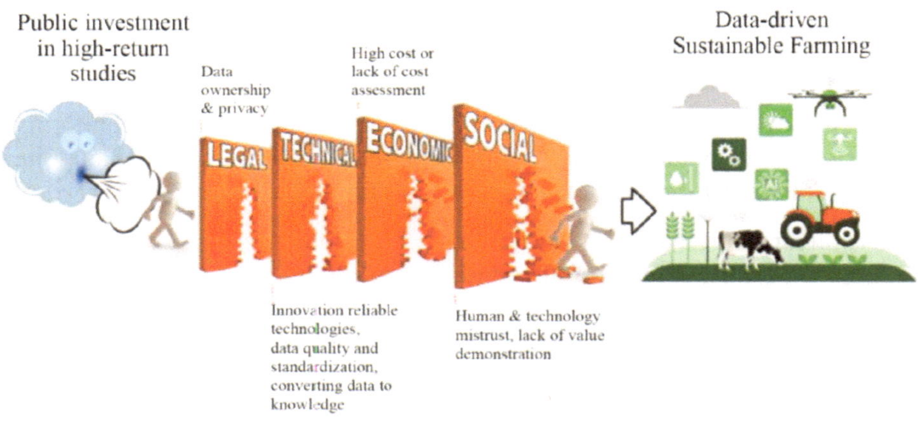

Fig. 3: Data driven farming for sustainable agriculture

Investing in data-driven farming holds immense promise for enhancing agricultural production and productivity. It serves multiple purposes, including adapting to and mitigating the impacts of climate change, optimising natural resource utilization, reducing risks, and bolstering resilience in farming practices. Ultimately, it contributes significantly to global food and nutrition security by fostering more efficient and productive agricultural systems.

Data-driven farming integrates big data to complement on-farm precision agriculture by leveraging farm-specific data at the right time and in the right manner to make informed decisions. Its application is being explored and tested extensively in many developing countries to determine where the highest agricultural yields from data utilization can be achieved. However, while data-driven farming offers vast potential, it's not equally accessible to smallholder farmers in developing countries. These farmers encounter significant challenges and risks in ensuring that their digital investments translate into tangible benefits. Two primary hurdles stand out: firstly, gaining access to pertinent data and services offered by external sources, and secondly, safeguarding their data sharing to avoid compromising their positions (Ranjan *et al.*, 2022b).

The opportunities, challenges, and risks in data-driven farming vary across various dimensions. Localised and imported data, when tailored precisely to on-farm situations, can provide substantial advantages (Zhang, 2016). Data shared by fellow farmers locally and globally should ideally result in more relevant services and products for individual farmers. Localised data, limited to small geographic areas, generally pose fewer challenges since it either directly pertains to their farms or lies entirely outside their operational scope. Imported data, however, presents challenges related to availability, accessibility, and usability. External data and services need adaptation to align with farmers' needs and capacities to be valuable. Exported data, involving sharing, carries both risks and benefits. Farmers must take precautions to avoid exploitation of the data they share, necessitating additional measures to ensure that shared data respects their interests and doesn't compromise their positions (Nicoleta *et al.*, 2019).

Data-driven farming, an integral facet of precision agriculture, leverages technology and data analytics to revolutionise decision-making in agriculture. It involves collecting, analyzing, and utilising vast amounts of agricultural data to enhance farm operations, boost productivity, and foster sustainability. Here's a detailed breakdown:

1. Data Collection

- Remote Sensing and Imaging: Utilises satellite imagery, drones, and remote sensing technology to capture soil health, crop growth, pest presence, and weather patterns.
- On-field Sensors: Devices gather real-time data on crucial factors like soil moisture, nutrient levels, and temperature, influencing crop growth.
- Machinery Telemetry: Modern farming equipment equipped with sensors records operational data such as seeding depth, fertiliser usage, and harvesting yields.

2. Data Integration and Analysis

- GIS and Data Platforms: Geographic Information Systems integrate diverse data sources, allowing farmers to visualise and analyse field information spatially.
- Big Data Analytics: Sophisticated algorithms process extensive datasets, identifying patterns and anomalies to derive actionable insights for farmers.

3. Decision Support Systems (DSS)

- Predictive Models: Machine learning algorithms use historical data to forecast crop yields, disease outbreaks, and optimal planting times.
- Prescriptive Analytics: DSS provides actionable recommendations like precise irrigation schedules and tailored fertiliser applications based on analysed data.

4. Precision Resource Application

- Variable Rate Technology (VRT): Informed by data analysis, VRT enables precise application of resources like fertilisers and pesticides tailored to specific field areas.
- Auto-steering and Guidance Systems: GPS-guided equipment ensures accurate input deployment, reducing overlap and waste.

5. Monitoring and Feedback Loop

- Yield Monitoring: Track crop yields during harvesting, linking them to specific field areas for performance evaluation.
- Continuous Monitoring: Real-time monitoring allows immediate adjustments, aligning farming practices with changing conditions.

6. Environmental Sustainability

- Emphasises sustainable practices by optimising resource usage, reducing chemical inputs, curbing water wastage, and preventing soil degradation.

7. Farm Management and Planning

Informed decision-making aids in long-term planning, including crop rotation, land optimization, and risk mitigation tied to climate change.

8. Adoption of Advanced Technologies

Integrating AI, IoT, and machine learning empowers farmers to continually refine data-driven strategies for enhanced sustainability and efficiency.

Data-driven farming empowers farmers to make informed decisions, optimise resource allocation, minimise waste, and increase productivity while ensuring sustainable agricultural practices. By harnessing technology and data analytics, it facilitates an efficient, resilient, and environmentally conscious approach to farming. Data-driven farming presents promising opportunities but also unique challenges, particularly for smallholders in developing countries who face hurdles in accessing and using relevant data. Tailoring data to specific farm needs is crucial, but precautions are needed in data sourcing, sharing, and ethics to protect farmers.

Modern farming practices generate vast data through advanced technologies, initially overwhelming but valuable when processed correctly. Agricultural technology providers develop data analytics for farm applications, benefitting not just growers but also industries like insurance and commodity markets. Standardisation is key for efficient data sharing, ideally supported by a common infrastructure. Emerging technologies bring unique challenges, especially in open platforms, necessitating attention to data ownership, privacy, and security. Balancing innovation with regulations is crucial.

Modern Farming Techniques

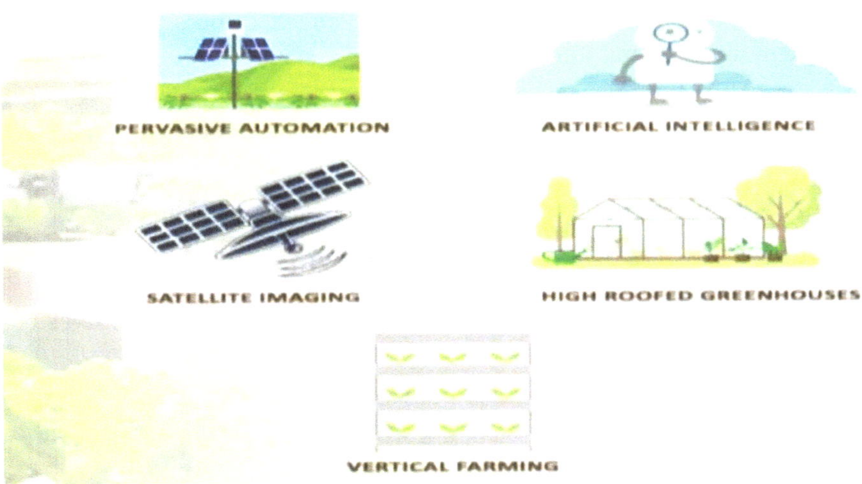

Fig. 4: Modern farming techniques

Smaller farmers could benefit from public institutions' open-source analytics, potentially facilitated by initiatives from entities like USDA and NSF. As agriculture trends towards automation and data-driven decisions, IoT adoption defines the future. The value of raw farm data increases when combined with effective algorithms. Agricultural Big Data processing is evolving but requires collaboration between data scientists and agricultural experts. Establishing shared platforms is crucial, with universities playing a pivotal role in providing data utilities.

The central idea behind data-driven precision agriculture is to enhance the efficiency, productivity, and sustainability of farming practices by collecting, analysing, and utilising vast amounts of data to make precise and informed decisions at every stage of the agricultural process. Here's a breakdown of the key components of this concept:

1. Data Collection: Gathering diverse data—soil info, weather, crop and livestock specifics—via technologies like sensors, drones, and satellites.
2. Data Analysis: Processing collected data with advanced algorithms to unveil valuable insights and patterns, enabling a better understanding of farm conditions.
3. Precision Decision-Making: Armed with deep insights, farmers make granular decisions, optimising resource use (water, fertilisers) based on real-time needs, cutting waste.
4. Customization: Tailoring approaches to specific field sections or individual plants/animals, factoring in soil, weather, and more for maximum yield and minimal impact.
5. Sustainability: Promoting eco-friendly practices by reducing resource overuse, minimising environmental effects, and enhancing land and water utilization for long-term viability.
6. Profitability: Boosting profits by cutting costs, improving yields, and making informed decisions about planting, harvesting, and marketing.
7. Risk Management: Identifying and mitigating risks early—disease outbreaks, pests, adverse weather—to safeguard crops and livestock through timely interventions.
8. Continuous Improvement: An ongoing process involving data collection, analysis, and adjustments, fostering continuous learning and adaptation for improved practices.

In essence, data-driven precision agriculture harnesses data and tech to optimise farming for efficiency, sustainability, and profitability. It represents a transformative shift in agriculture, making it more data-centric and informed.

Conclusion

Advancing in data-driven farming promises a revolution in agriculture, elevating production, productivity, and sustainability. Its potential spans mitigating climate change effects, efficient resource use, risk reduction, and bolstering farming resilience. Crucially, it significantly contributes to global food security by optimising decision-making using big data. However, this shift isn't equally accessible to smallholders in developing countries, facing challenges in accessing relevant external data and services without compromising their positions. Tailoring data to specific farm needs can benefit farmers but requires careful consideration of data sources, nature, and sharing to safeguard their interests. Data, pivotal in modern agricultural tech systems, serves as both an essential input and valuable product. Open data holds empowering potential but demands adherence to clear principles, rules, and ethics to prevent vulnerabilities in precision agriculture. Modern farming generates vast, valuable data sets, benefiting not only growers but also industries like insurance and commodities. Standardisation and a shared infrastructure for data sharing are vital for maximising data potential. As technology progresses, addressing data ownership, privacy, and security challenges is essential, especially with open platforms promising accelerated innovation. Striking a balance between innovation and protection is crucial for the sustainable growth of data-driven farming in agriculture.

References

Mohapatra, P., Seleiman, M.F., Mandal, R., Pramanik, K., Maity, T.K., Tarafdar, J., *et al.* (2023). Efficiency of RAPD and SSR markers in assessing genetic diversity in summer onion (Allium cepa L.) genotypes. Notulae Botanicae Horti Agrobotanici Cluj-Napoca, 51(3), 13369.

Nicoleta, T., Stavros, S., and Manos, R. (2019). Data-Driven Decision Making in Precision Agriculture: The Rise of Big Data in Agricultural Systems. Journal of Agricultural & Food Information 20(4), 344-380.

Ranjan, S., Sow, S., Kumar, S., Ghosh, M., Roy, D.K., Dutta, S.K., and Pramanick, B. (2022b). Enhancing Nutrient Use Efficiency and Productivity of Cereals through Site Specific Nutrient Management. Journal of Cereal Research, 14(3), 229-242.

Ranjan, S., Sow, S., Kumari, J., Kumar, N., Roy, D.K., and Pramanick, B. (2022a). Use of Remote Sensing in Agriculture. Agriculture Letters, 3(9), 48-52.

Sow, S., Singh, G., Ghosh, M., Dutta, S.K., Mandal, N., Kumar, S., and Ranjan, S. (2023). Nitrogen-management strategy through leaf-colour chart and SPAD meter for optimising the productivity in irrigated wheat (Triticum aestivum). Indian Journal of Agronomy, 68 (2), 219-222.

Zhang, Q. (2016). Precision agriculture technology for crop farming. CRC Press Taylor & Francis Group, Broken Sound Parkway NW, Suite, Boca Raton, FL-33487-2742.

13

Recent Advances in Management of Problematic Soil

***Reena Kujur*[1], *Shashi Kant Ekka*[2], *Heba I. Khan*[1] *Eugenia P. Lal*[1] *and Ram Bharose*[3]**

[1]*Department of Biological Sciences, SHUATS, Uttar Pradesh*

[2]*Department of Horticulture, SHUATS, Uttar Pradesh*

[3]*Department of Soil Science and Agricultural Chemistry, SHUATS, Uttar Pradesh.*

Abstract

The majority of regenerative farming practices are becoming increasingly recognised for their ability to enhance soil health and mitigate climate change. Increased soil organic matter, carbon sequestration, and improved water retention are all made possible by crop rotation, cover crops, and reduced tillage. Traditional agricultural methods have often led to soil degradation, reduced fertility, and increased reliance on chemical inputs, contributing to environmental degradation and threatening long-term food security. To counteract these trends, a paradigm shift towards sustainable practices is essential to manage problematic soil. This chapter addresses the soil-related challenges, encompassing soil health management, precision agriculture, soil amendments, remediation techniques, conservation practices, and digital technologies. This enables farmers to optimise resource use, reduce environmental impact, and enhance overall productivity.

Keywords: *Organic matter, Precision agriculture, Problematic soil, Soil degradation*

Introduction

The importance of soil in agriculture cannot be overstated, as it serves as the foundation for crop growth and sustenance. However, many regions grapple with problematic soils characterised by issues such as poor fertility, erosion, contamination, and structural degradation. In recent years, significant steps have been made in addressing these challenges through innovative approaches and technologies. This change in outlook includes the combination of imaginative advancements, protection procedures, and biological standards to enhance farming efficiency while limiting ecological effects (Singh *et al.*, 2011).

Farmers are challenged with the management of problematic soils characterised by fertility issues, erosion, contamination, and structural degradation. Sustainable agricultural techniques depend on the efficient management of problematic soils, which is also critical to environmental sustainability, global food security, and farming systems' ability to withstand climate change. The idea of changing soil and harvest management practices a complex system that incorporates accurate horticulture, agroecology, and computerised cultivating innovations (Fu *et al.*, 2021).

The main objective of this chapter is to provide a comprehensive exploration of the recent advances in the management of problematic soils, highlighting the innovative strategies and technologies that have emerged to address these challenges. As we delve into each aspect of soil management, it becomes apparent that a nuanced and integrated approach is essential for tackling the complexities associated with problematic soils. By examining these recent developments, we gain insights into how modern agriculture is adapting to ensure the sustainable use of soil resources while striving for increased agricultural productivity and environmental stewardship. High salinity and heavy metal contamination in problematic soils have an impact on ecosystem services associated with the soil. Bioremediation of these soils ensures plant–soil health for sustainable development on a global scale (Raj *et al.*, 2022).

Problematic Soil

Problematic soils refer to soils that present challenges or limitations for certain land uses or agricultural activities due to their specific characteristics such as the presence of soluble salts, soil reaction, exchangeable sodium percentage, waterlogging, aeration etc. These characteristics can adversely affect the optimum soil productivity and are called problematic soils (Ashokkumar, 2019). They can negatively impact plant growth, water movement, and various other soil functions. Problematic soil can be classified into various forms which are as given in Table 1.

Table 1. Classification and characteristics of problematic soil in India

Problematic soils	**Characteristics**
Nutrient-Deficient Soils	Soils lacking essential nutrients like nitrogen, phosphorus, or potassium can limit plant growth.
Acid soils	Soil pH is less than 6.5, Fe, Al toxicity (Strong acid soil).
Salt affected soils a. Saline soils	Electrical conductivity is greater than 4.0 dS/m, High osmotic potential and nutrient imbalance.
b. Sodic soils	Exchangeable sodium percentage is greater than 15, Deteriorated physical condition, Na toxicity, nutrient imbalance

c. Saline sodic soils	Electrical conductivity is greater than 4.0 dS/m and ESP is greater than 15, High osmotic potential, deteriorated physical condition, nutrient imbalance
Alkaline Soils	Alkaline soils have high pH levels, which can also affect nutrient availability. Some plants are sensitive to high soil pH, leading to nutrient deficiencies.
Contaminated Soils	Soils contaminated with pollutants, such as heavy metals, pesticides, or industrial chemicals, can pose risks to human health and the environment.
Eroded Soils	Soil erosion removes the top layer of fertile soil, which can result in reduced fertility and decreased water-holding capacity. Erosion is often caused by factors such as water runoff and wind.

Soil Management

There are four main components of soil: air, water, mineral particles, and organic matter. The organic matter can be further classified as organisms, roots, and humus. There are a number of organic and inorganic substances dissolved in the soil pore water as soil solution, which are immediately available to plants.

Organic matter is a complex of living organisms and materials that have decomposed in various stages over time. The availability of nutrients depends on the type of organic matter and the activity of soil organisms. Decomposition is the process of breaking down organic matter into simpler organic and inorganic compounds which is managed by soil microorganisms (Ismail-Meyer *et al.*, 2018).

- Efficiency: The reappearance obtained from sustainable land management is more than just estimating financial profits; it is evaluated to include the benefits that will be obtained from the protective, health and aesthetic purposes of land use (Mir *et al.*, 2022).
- Security: The management models that support the stability between land use and the existing environmental conditions reduce production threats,. However only those approaches that emphasise commercial anxiety increase the risk (Tuğrul, 2019).
- Protection: Soil and water resources should be taken under strict protection for future generations (Young, 2000). There may be additional protection priorities on a local level, such as maintaining genetic diversity or protecting specific species of plants and animals.
- Vitality: Uses of usable land that don't match local conditions cannot survive.

- Acceptability: If land use methods have a negative social effect, they are likely to fail over time. Occasionally, social and economic impacts do not have a positive impact on those directly affected.

Method of Soil Reclamation

i. Mechanical Method

Mechanical methods of soil reclamation involve using various equipment and machinery to physically modify and improve the structure and properties of the soil. These methods are employed to address issues such as compaction, erosion, and poor drainage. Some common mechanical methods of soil reclamation are:

- Contour Ploughing: Plowing along the contours of the land helps reduce water runoff and soil erosion. This method helps to slow down water movement and allows it to infiltrate into the soil.
- Deep Ploughing: This involves breaking up compacted layers of soil by using plows, that can penetrate to greater depths. Deep plowing helps improve aeration, water infiltration, and root penetration.
- Ditching and Drainage: Installing drainage systems, such as ditches and tiles, can help improve the drainage of waterlogged soils. This prevents water from accumulating on the surface, reducing the risk of waterlogging.

ii. Hydrological Method

- Flushing: Flushing, or the process of washing salts off the surface using runoff water that collects at the lower end of the field, is one way to get rid of salts. However, flushing cannot remove soluble salts through the soil; instead, the salts must rise to the surface in order to be eliminated. This is a significant disadvantage of the method. The flushing method is employed in areas where moisture distribution features in the soil profile are extremely poor.
- Leaching: Leaching is the process of dissolving and transporting soluble salts by the downward movement of water through the soil. There are two ways to accomplish leaching: (i) floods and (ii) sprinkler techniques. The flooding process eliminates the salt at a deeper level and uses more water. It has been discovered that the sprinkler approach effectively uses less water to remove salts from the top layer (60 cm).
- Drainage: The practice of draining surplus water from the soil is known as drainage in agriculture. Excess water discharged by flow over the soil surface is referred to as surface drainage, and flow through the soil is termed internal or subsurface drainage.

iii. Chemical Method

The exchange complex is partially saturated with Na in both saline and sodic soils. The reclamation procedure in such cases also involves the use of amendments for replacing exchangeable Na, which involves:

Soluble calcium salts- Calcium chloride, Gypsum, Calcium sulphate.

Acid or acid formers – Sulphur, Iron sulphate, Aluminium sulphate, Pyrites etc.

iv. Organic amendments

Application of organic compounds primarily improves the physical condition of the soil by enhancing soil aggregation. The most common organic amendment is farmyard manure (FYM) which is added in the first year of reclamation @ 50 tonnes/ha and is reduced to half in succeeding years. The efficiency of gypsum has been found to increase when applied along with FYM.

Bioremediation of Problematic Soil

Bioremediation is an environmentally friendly and cost-effective method for addressing various soil-related issues, especially those caused by contaminants. This approach relies on the use of living organisms, such as bacteria, fungi, and plants, to break down or neutralize pollutants in the soil. Here are some key strategies for bioremediation of problematic soils:

Phytoremediation: Certain plants, known as hyperaccumulators, can absorb and accumulate heavy metals from the soil. Once the plants have accumulated the contaminants, they can be harvested and removed, effectively reducing the pollutant levels in the soil (Ranjan and Sow, 2021).

Biostimulation: Adding nutrients such as nitrogen, phosphorus, and other essential elements to the soil can enhance the growth and activity of indigenous microorganisms. This stimulates their ability to break down contaminants.

Aeration: Improving soil aeration helps create an aerobic environment, which is favorable for the growth of aerobic bacteria that can degrade organic pollutants.

Bioaugmentation: Inoculation of specific strains of bacteria, fungi, or other microorganisms that are capable of degrading or transforming contaminants can accelerate the bioremediation process.

Mycoremediation: Some fungi have the ability to break down and absorb contaminants. It involves using fungi to degrade pollutants in the soil. For example, certain species of fungi can break down hydrocarbons.

Bioventing: Aerobic Soil Venting: Bioventing involves the introduction of air into the soil to stimulate the activity of aerobic microorganisms. This process can enhance the natural degradation of volatile organic compounds (VOCs) in the soil.

Constructed Wetlands: Constructed wetlands use plants to treat contaminated water. The plants absorb and accumulate pollutants from the water, providing a natural filtration mechanism.

Revolutionizing Management Practices for Sustainable Agriculture

By 2050, crop production may not be able to keep up with the huge demand brought on by population growth (Ray *et al.*, 2013). It is therefore necessary to employ a variety of management techniques in order to increase crop productivity. As we move forward, it is important to keep in mind that agriculture and the earth's future are interdependent. In order to increase plant yield and health, following these management practices is crucial.

Artificial Intelligence and Machine Learning

A new wave of precise and intelligent agricultural methods is being brought about by Artificial Intelligence (AI) and Machine Learning (ML), revolutionising crop management. With the use of sophisticated algorithms, AI systems are able to anticipate crop health, weather patterns, and soil conditions, among other factors, with a high degree of accuracy. By using their predictive power, farmers may minimize waste and maximise resource efficiency by choosing wisely when to plant, water, and harvest. Higher yields and better crop quality result from these automated procedures, which also save time and effort and reduce human error (Shaikh *et al.*, 2022).

Furthermore, early diagnosis of diseases or insect infestations is made feasible by the machine learning models' ability to recognise trends in the growth and health of plants. Prompt detection and intervention can avert extensive harm, guaranteeing maximum yields. Agriculture's future can be improved by utilising AI and machine learning to guide farming toward more profitability, sustainability, and consistency.

Drones Practices in Agriculture

By providing a fresh viewpoint on sustainable farming, drones are also transforming the agriculture industry. Drones with excellent imaging equipment can capture comprehensive aerial photos that provide detailed information about soil and crop health with reduced impact on the environment. With this aerial perspective, farmers can identify problems like insect invasions or nutritional deficits at an early stage. Early detection of these issues enables prompt intervention, minimising crop losses and guaranteeing ideal yields (Bolman, 2016).

Precision Farming: Increasing Productivity, Reducing Waste

Sustainability is being greatly advanced by precision agriculture, an innovative farming method that uses technology to maximise crop yield and resource efficiency. Precision agriculture enables farmers to manage their fields with a previously unheard-of degree of accuracy and detail thanks to advancements like GPS mapping, remote sensing, and data analytics (Sow and Kumar, 2023). By using this information, farmers can better understand the spatial variability of their fields and adjust their farming practices to suit the unique requirements of various regions. This reduces waste and costs by making better use of inputs like water, seeds, and fertilisers (Ranjan *et al.*, 2022). They promote the monitoring of weather patterns, soil moisture content, and crop health—elements that have a big impact on crop yield. These technologies are enhanced by data analytics, which processes the massive volumes of field data collected. They offer strategic insights into planting schedules, irrigation requirements, and pest management strategies through predictive modeling and trend analysis (Karunathilake *et al.*, 2023).

Challenges and Opportunities

As agriculture moves towards becoming more sustainable, it faces difficult obstacles from competition for natural resources like land, water, and biodiversity to socioeconomic concerns about the growth of communities, rural livelihoods, emerging markets, and environmental integrity in light of climate change (FAO 2016; Hall and Clark 2010). The management of problematic soils is fraught with challenges arising from the inherent heterogeneity of soil conditions, financial constraints, and the long-term sustainability of adopted practices. Farmers and landowners face difficulties in accessing accurate soil data, hindering the implementation of tailored strategies. Resistance to change, often rooted in traditional practices, coupled with a lack of awareness and education about sustainable soil management, further complicates effective interventions. To overcome these challenges, there is a need for integrated, context-specific, and science-informed approaches, as well as collaborative efforts among stakeholders, policymakers, and researchers to promote sustainable and resilient soil management practices.

Sustainable agriculture needs to be encouraged as we address the issues brought on by climate change and expanding food demand in order to meet the changing needs of a growing global population while minimising environmental damage (Chouhan *et al.*, 2021).

Conclusion

Recent advances in the management of problematic soils have showcased promising developments, leveraging innovative technologies and holistic approaches. Precision agriculture, guided by data-driven insights from advanced sensing technologies and satellite imagery, allows for targeted interventions tailored to specific soil conditions. The integration of artificial intelligence and machine learning in soil management systems has improved predictive modeling and decision-making processes. Continued research, education, and collaboration are essential to further refine and implement these innovative approaches for the benefit of agriculture, ecosystems, and global sustainability.

References

Ashokkumar V. (2019). Problematic soils and their management, Conference: UG Teaching Junagadh Agricultural University.

Bolman, B. (2016). A revolution in agricultural affairs: dronoculture, precision, capital. In The Good Drone. Routledge, pp. 129-152.

Chouhan, G. K., Verma, J. P., Jaiswal, D. K., Mukherjee, A., Singh, S., de Araujo Pereira, A. P., *et al.* (2021). Phytomicrobiome for promoting sustainable agriculture and food security: Opportunities, challenges, and solutions. Microbiological Research 248, 126763.

FAO. (2016). Food and Agriculture Organisation, The State of Food and Agriculture 2016 (SOFA): Climate Change, Agriculture and Food Security; Food and Agriculture Organisation of the United Nations: Rome, Italy, 2016.

Fu, X., Zheng, Z., Sha, Z., Cao, H., Yuan, Q., Yu, H., and Li, Q. (2022). Biorefining waste into nanobiotechnologies can revolutionise sustainable agriculture. Trends in Biotechnology, 40(12), 1503-1518.

Hall, A., and Clark, N. (2010). What do complex adaptive systems look like and what are the implications for innovation policy? J. Int. Dev., 22, 308–324.

Ismail-Meyer, K., Stolt, M. H., and Lindbo, D. L. (2018). Soil organic matter. In Interpretation of micromorphological features of soils and regoliths, Elsevier, pp. 471-512.

Karunathilake, E. M. B. M., Le, A. T., Heo, S., Chung, Y. S., and Mansoor, S. (2023). The path to smart farming: Innovations and opportunities in precision agriculture. Agriculture, 13(8), 1593.

Ranjan, S., Sow, S., Kumari, J., Kumar, N., Roy, D.K., and Pramanick, B. (2022). Use of Remote Sensing in Agriculture. Agriculture Letters, 3(9), 48-52.

Ranjan, S., and Sow, S. (2021). Phytoremediation: An eco-friendly approach towards clean and green future. The Pharma Innovation Journal, 10(3), 839-850.

Raj, A., Jhariya, M. K., Khan, N., and Banerjee, A. (2022). Bioremediation of Problematic Soil for Sustainability. Microbial Biotechnology: Role in Ecological Sustainability and Research, 201-223.

Ray, D.K., Mueller, N.D., West, P.C., and Foley, J.A. (2013). Yield trends are insufficient to double global crop production by 2050. PLoS ONE, 8, e66428.

Shaikh, T. A., Rasool, T., and Lone, F. R. (2022). Towards leveraging the role of machine learning and artificial intelligence in precision agriculture and smart farming. Computers and Electronics in Agriculture, 198, 107119.

Singh, J. S., Pandey, V. C., and Singh, D. P. (2011). Efficient soil microorganisms: a new dimension for sustainable agriculture and environmental development. Agriculture, Ecosystems & Environment, 140(3-4), 339-353.

Sow, S., and Kumar, N. (2023). Application of remote sensing in sugarcane cultivation. In: Sustainable Sugarcane Production and Utilisation: Issues and Initiatives, Dr. Rajendra Prasad Central Agricultural University, Pusa, Bihar, pp 202-206.

Young, A. (2000). Land resources: now and for the future. Cambridge University Press.

14

Recent Innovations in Nutrient Management for Maximising Crop Production

Souvik Sadhu[1], Rajni Prabha Rani[1], Koushik Barik[2] Dinabandhu Samanta[3] and Ritwik Sahoo[4]

[1]Department of Soil Science and Agricultural Chemistry, Bihar Agricultural University Sabour, Bihar

[2]Department of Agronomy, UBKV, Pundibari, Cooch Behar, West Bengal

[3]Department of Pomology and Post-Harvest Technology, UBKV, Pundibari Cooch Behar, West Bengal

[4]Department of Plant Pathology, UBKV, Pundibari, Cooch Behar, West Bengal

---

Abstract

Proper nutrient management is a vital need in today's context in order to maintain a balance between the food requirements for the ever-increasing global population and rapidly disappearing cultivable lands. The so called traditional nutrient management practices can no longer provide an adequate amount of food of proper quality for the sustenance of global residents as the fertility of the land has been severely degraded due to the vigorous removal of nutrients from the soil by high-yielding varieties and not returning to the soil what is being taken from it. Moreover, indiscriminate use of synthetic fertilisers without considering organics as a diet of soil forcibly alters the physical, chemical and biological equilibria of soil as well as severely deteriorates the nutritional profile of the food products. This ultimately results in malnutrition and hunger thereby collapsing the sustainable development goals (SDGs). Innovative nutrient management practices like nano fertiliser formulations, biofertilisers and nutrient mobilising microbes, slow-release fertilisers, foliar application, seed and seedling priming, seaweed extract etc. are being used nowadays to combat the dual problem of crop yield maximisation and boosting the nutritional profile of the food products. In this chapter we will discuss some innovative nutrient management strategies and how they can be a trustworthy option to combat the above-mentioned problems.

***Keywords**: Nutrient management, SDGs, Synthetic fertiliser, Yield*

Introduction

According to the United Nations (UN, DESA 2019) the world population will reach around 9.7 billion by 2050. The Food and Agriculture Organisation (FAO) has predicted that the world food grain production must increase by 70% by the year 2050 to feed this ever-growing population (Bernard *et al.*, 2017). United Nations World Population Prospects (2022) also predicted the same (Fig 1). However, it is a matter of great concern that the global food grain production has not increased to any significant level (Fig 2) in the past 15 years. As seen in Fig 2, the global food grain production trend has reached a plateau. The main reason behind this fact is the degradation of this valuable natural resource (soil) beyond the tolerance limit due to anthropogenic activities such as imbalanced fertiliser use with the application being largely skewed in favour of nitrogen (N) fertilisers. As per the available data from 2020-21, in nutrient terms, India consumes a total of 32.5 million tonnes (Mt) of fertilisers swhich are distributed over a stable net cultivated area of 141 million hectares (Mha) (Singh *et al.*, 2022). Associated with this is the expenditure of a huge amount of energy for producing these fertilisers, which principally comes from fossil fuel burning, pumping a gigantic quantity of groundwater to solubilise these fertilisers and so on. Not only nitrogen, but the usage of phosphatic and potassic fertilisers has also witnessed an upsurge in the last 50 years (Fig 3).

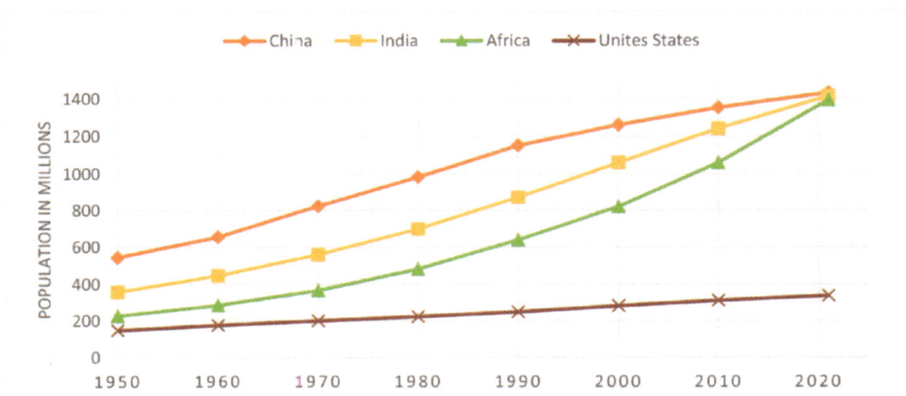

Fig. 1: Global population trend

Data Source: United Nations, World Population Prospects (2022)

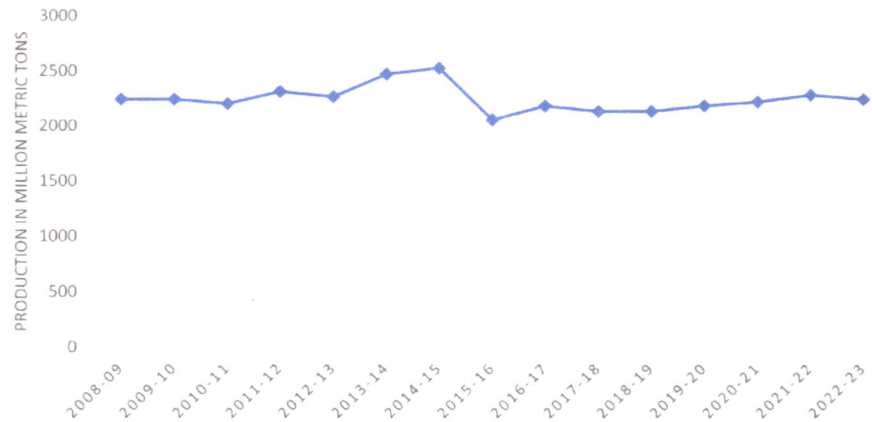

Fig. 2: Total global grain production from 2008/2009 to 2022/2023

Source: Statista 2023.

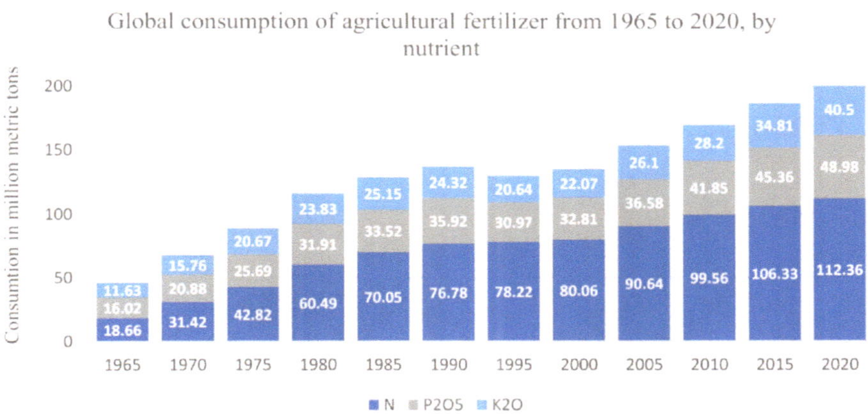

Fig. 3: Trend of global fertiliser consumption (over the years)

Source: Statista 2023

More concerning is that farmers apply these fertilisers in the soil irrationally without knowing the factors that control their phyto-availability with a belief that the more fertiliser they apply, the greater the yield they will achieve. This is the reason why the nutrient use efficiency of applied fertilisers is very low only 30-35% for nitrogen (Congreves *et al.*, 2021), 10-15% for phosphorus (Roberts *et al.*, 2015), 30-50% for potassium and 1-2% for micronutrients (Fageria *et al.*, 2014). The inevitable consequences of these faulty nutrient management practices are declining factor productivity, a declining ground water table, deterioration in soil fertility, air and ground water pollution, and stagnating farm incomes. Today agriculture and the allied sector contribute to about 24% of the global greenhouse gas emissions (Fig 4). Fertiliser s are one of the most crucial agricultural inputs to increase crop productivity (Osman *et al.*, 2020). It is not that poor fertilisation schedule is solely responsible for crop yield stagnation. Focusing less on soil physical condition, poor water management etc. are also contributors to the distressing outcome. However, a careful outlook will reveal that all these factors are in one way or other related to nutrient management (Fig 5). India will need 405 Mt of food grains to feed 1.68 billion people by the year 2050 from the same or less arable area (Singh *et al.*, 2022). Unanimously, the world agriculture will face much challenges to accommodate this large population with foodgrains (Sadhu *et al.*, 2023). It has been predicted that if fertiliser nitrogen use efficiency can be increased by 10 %, (from 40 to 50 % on an average) it can exempt the consumption of 1.7 Mt of fertiliser nitrogen equivalent to Rs. 2000 crores per year. At the same time, it can minimise environmental footprints of 22000 tonnes of CO2 equivalent each year (New Delhi Soil Health Declaration 2015).

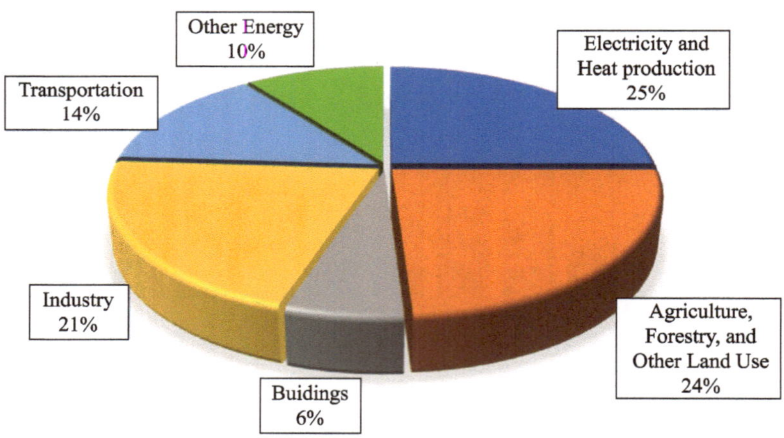

Fig. 4: Global greenhouse gas emission by economic sector

Source: IPCC (2014)

Fig. 5: How better management is related to other factors of crop production

Traditional Nutrient Management Practices and Their Shortcomings

Conventional agriculture systems are described as the predominant farming practices and systems of crop production adapted by farmers in a particular region. It is defined as being based on the rigorous usage of agrochemicals with the aim of maximising agricultural crop yields, encompassing extensive tillage to manage the physical traits of the soil and restrict the weed population, mono-cropping and poor recycling of resources (Giller *et al.*, 2017). Agriculture can be classified as conventional if:

- It alters the natural environment
- Synthetic chemicals are used to maintain plants
- A significant amount of chemical and energy input is required to produce the highest possible yield of crops
- Deteriorates soil quality
- Eliminates biodiversity.

So, in this context there is an urgent need to think about resource conserving and more holistic approach of crop nutrient management which not only considers the soil as a crop producing medium, but considers it as a natural body and at the same time do not force climate change and its degradation.

In the last two decades, with gradual advancement of awareness among the scientists about the fertiliser use efficiency and the environmentalists about the environmental hazards these unutilised fertilisers cause, several innovative fertiliser management practices have been developed. Some of these are nano fertiliser formulations, biofertilisers and nutrient mobilising microbes, slow-release fertilisers, seaweeds etc.

Few Case Studies About The Success of Some Novel Nutrient Management Practices

Nano Fertiliser Formulations

Nanotechnology deals with "the understanding and manipulation of matter at dimensions of roughly 1-100 nm, where unique properties make novel applications possible" (Nanoscale Science Energy and Technology Subcommittee 2007). With their high surface area to volume ratio, regulated release kinetics, and high sorption aptitude nano particles (NPs) possess high reactivity, their nano-scale size enables plants to absorb them efficiently through stomata (Uzu *et al.*, 2010) and ultimately, they can be used as a novel nutrient delivery system. Various nano fertiliser formulations like Manganese (Dimpka *et al.*, 2018), Zinc (Adhikary *et al.*, 2020) and iron oxide NPs (Rokana *et al.*, 2023), nano hydroxyapatite (Elsayed *et al.*, 2022), nano clay-polymer composites (Sarkar *et al.*, 2014; Mandal *et al.*, 2018), carbon nanotube (Mondal *et al.*, 2011) have been successfully synthesised and evaluated. Lu *et al.* (2002) demonstrated that with the mixed application of SiO2 and TiO2 NPs soybean reported enhanced fertiliser as well as water uptake. NPs application also resulted in higher nitrate reductase (responsible for biological N2 fixation) and antioxidant activity. Enhanced rate of seed germination and yield parameters in the mustard crop have been reported by Mondal *et al.* (2011) with the use of multi-walled carbon nanotubes (MWCNTs). The study also revealed that oxidised MWCNTs were having better stimulative effect than the non-oxidised counterparts. Similar results with the application of MWCNTs were also reported by Khodakovskaya *et al.* (2009, 2012) for tomato and tobacco. Foliar spray of nano urea along with nano Zn and Cu significantly increased plant height, number of cobs per plant, number of seeds per cob, cob length and test weight in maize (Singh *et al.*, 2022). Upadhayay *et al.* (2023) opined that up to 25% of the recommended dose of nitrogen (RDN) could be curtailed without any yield penalty if 50% of the top-dressed N is accompanied by 2 additional foliar spray of nano urea in the maize-wheat and pearl millet-mustard cropping system. Saurabh *et al.* (2019) documented that when urea particles were entrapped within a nano clay polymer composite (nitrogen nano clay polymer composite or N-NCPC) and used as a nitrogen fertiliser, it could lower down the N2O emissions from wheat and rice rhizosphere to the tune of 12.4% and 16.1% respectively when compared with sole urea application. Regulated release of mineral N with better synchronisation with crop demand in the case of N-NCPC was described to be the responsible factor behind this phenomenon. Various studies revealed better uptake of NPs into plant cells due to their tendency to be attached with carrier proteins and membrane transporters located in the roots such as aquaporins (Nair *et al.*, 2010) and also forming complexes with root exudates (Kurepa *et al.*, 2010).

Biofertilisers and Nutrient Solubilizing Microbes

Biofertiliser is a substance containing living microorganisms that, when applied to seeds, plants or soil, colonize the rhizosphere or internal parts of plants and promote plant growth by increasing the supply of nutrients to the host plant (Bardi and Malusà 2012; Malusa and Vassilev 2014). Biofertilisers are extensively used to speed up microbial processes that increase the availability of easily absorbable nutrients to plants. They improve soil fertility by capturing nitrogen from the atmosphere, dissolving insoluble phosphates and by producing plant growth promoting substances (Mazid and Khan 2015). Being environment friendly and cost effective in nature, their long-term augments soil fertility and promotes sustainability. On an average, there is a 10-40 % increase in crop yield due to biofertiliser application. has also been reported due to their N- fixation, P and K solubilisation, and increase in the quantity of protein, vitamin and essential amino acid (Bhardwaj *et al.*, 2014). Table 1 shows the various supportive roles biofertiliser s play in promoting plant growth and as a biocontrol agent. But certain flaws are also there in using these biological agents that restrict their scope of use. Their activity is very much dependent on the physio-chemical properties of the soil as well as environmental factors. Near-neutral soil reaction, loamy soil texture, and moderate to high organic carbon content favour their proper growth and multiplication. Adverse climatic conditions (extreme heat or cold and aridity), excessive use of agrochemicals, and poor soil aggregate stability are contrary and repugnant to their suitable working. Another component in this regard that often limits their adoption by the farming community is that the beneficial effect of biofertiliser application takes a considerable amount of time to reflect which is just a matter of 3-5 days in the case of chemical fertiliser s. So, farmers who want an instant outcome of the applied fertiliser are mostly skewed towards chemical fertilisers and reluctant towards biofertilisers.

Table 1. Biofertilisers in plant growth promotion and biocontrol (Mahanty *et al.*, 2016)

Sl. No	Biofertiliser type	Role in plant growth promotion	Role as biocontrol agent	Reference
1.	Pseudomonas fluorescens	Siderophore production	antifungal antibiotics production and inhibit phytopathogenic fungal growth	Kloepper *et al.* 1980; Nowak *et al.* 1994
2.	*Bradyrhizobium* sp Bradyrhizobium sp.	Siderophore production and phosphate solubilisation	HCN production	Afzal and Bano 2008; Deshwal *et al.* 2003
3.	Bradyrhizobium japonicum	same	Release of cell-wall degrading enzymes, inhibits fungal growth	Bardin *et al.* 2004; Afzal and Bano 2008; Shaharoona *et al.* 2006
4.	Azorhizobium	Nitrogen fixation	-	Sabry *et al.* 1997
5.	Streptomyces	IAA and siderophore production	Production of glucanases	Verma *et al.* 2010
6.	Arthrobacter	phosphate solubilisation	Chitinases synthesis	Bhattacharyya and Jha 2012
7.	Rhizobium meliloti	Siderophore production	Release of antibiotic and cell-wall degrading enzymes	Prabha *et al.* 2013; Bardin *et al.* 2004
8.	Frankia	Nitrogen fixation	-	Simonet *et al.* 1990
9.	Sphingomonas	Gybberelin synthesis	Competition	Khan *et al.* 2014

Seaweed Extract-based Fertilisers

In broad terms, seaweeds (originally developed by Milton, 1952) are masses of macroscopic marine algae belonging to the plant kingdom Thallophyta (Dhargalkar *et al.* 2001; Arioli *et al.* 2015). Researchers around the world have acknowledged seaweed as a potential source of plant nutrients. Compositional analysis of seaweed extracts from various taxonomic groups like green, red and brown algae has revealed that the extracts contain essential minerals such as K, Ca, Mg, Fe, Mn, Cu etc (Anantharaman *et al.* 2010; Tuhy *et al.* 2015). Moreover, a diverse array of organic compounds such as cellulose, protein, fibre, amino acids, and vitamins are abundantly present in this extract (Shevchenko *et al.*, 2007). This novel plant-based organic fertiliser has the potential to stimulate vital plant physiological attributes like seed germination, improved root growth, superior frost resistance, higher nutrient uptake as well as higher yield (Nabti *et al.* 2010). Besides supplying almost every essential nutrient element in balanced proportions, they also promote various physio-chemical properties of the soil such as moisture retention, soil structural stability (Dhargalkar and Pereira 2005), stimulate a wide range of microbial populations by modifying their microhabitat in the soil and provide a favourable environment for root growth and thereby nutrient mineralisation and mobilisation (Battacharya *et al.* 2015). A number of seaweed-based plant growth stimulants are available in the market (Table 2).

Table 2. Seaweed extract-based fertiliser available for commercial use (Khan *et al.*, 2009)

Product name	Seaweeds name	Applications
Agrokelp	Ascophyllum nododum	Plant growth stimulant
Kelpak	Durvillea antarctica	Plant growth stimulant
Sea winner	Durvillea potatorum	Plant biostimulatant
Sagarika (India)	Kappaphycus alvarezii	Plant growth promoter
Acadian	Ascophyllum nodosum	Plant growth stimulant
Algrin	Sargassum sp. + Ascophyllum nodosum + Laminaria sp.	Plant growth stimulant

The stimulative effect of this novel organic fertiliser has been established in several research experiments which have been summarised in Table 3.

Seaweed extract-based fertilisers not only have stimulative roles on crop yield and eco-physiological parameters, but also have been proven to be a soil ameliorant in various experiments. They have been shown to invigorate various soil enzymatic activities such as dehydrogenase, acid and alkaline phosphatase and microbial biomass.

Slow-release Fertilisers (SRFs)

SRFs have been defined as those fertilisers which either lag the nutrient availability post application thereby delaying their uptake and utilisation by plants or remain accessible to the plants for a considerably longer period compared to orthodox fertilisers (Ţolescu and Iovu 2010). These have been comprehensively studied worldwide to provide an innocuous, economical and proficient way of delivering nutrients due to their adeptness in delivering the essential nutrients at the target sites gradually, thereby minimising toxicity hazards (Gil *et al.*, 2020). At the same time the fertiliser dose can be cut down by 20-30% without any yield penalty, thus reducing the generation of harmful fertiliser by-products responsible for environmental pollution such as N2O, NH3, NO3- leaching (Wang *et al*, 2012; Cole *et al*, 2016). Patil *et al.* (2010) made a comparative study between conventional fertiliser (urea) at 100% RDN and coated fertiliser at 50 and 70% RDN. They found no significant differences among the treatments in terms of vegetative and yield parameters. Giving yield similar to the 100% RDN despite fertilised with lower doses of N highlights the higher NUE of the coated fertilisers. Yang *et al.* (2016) conducted an incubation trial in China with different sources of N to assess the release pattern of N from these sources.

Table 3. Experimental findings of seaweed extract application in various crops (Raghunandan *et al.*, 2019)

Sl. No	Seaweed	Crop	Effect	Reference
1.	Ascophyllum nodosum	Onion	Improved growth and yield parameters with low downy mildew incidence	Dogra and Mandradia (2012)
2.	Corallina elongata	Maize	Increased plant height, number of leaves and increase in K uptake	Safinaz and Ragaa (2013)
3.	Padina gymnoposra	Tomato	Better seed germination, greater plumule and radical length and seedling vigour	Rosalba Mireya Hernández-Herrera *et al.* (2014)
4.	Sargassum sp. + Ascophyllum nodosum + Laminaria sp.	Strawberry	Foliar application of seaweed-extract positively influenced various vegetative growth characteristics, Yield and nutritional quality of fruits as well as plants.	El-Miniawy *et al.* (2014)
5.	Stoechospermum marginatum	Brinjal	Higher fruit yield and fruit weight	Sivasangari Ramya *et al.* (2015)
6.	Codium tomentosum	Wheat	Increased germination, seedling growth, chlorophyll and carotenoids content	Mohy El-Din (2015)
7.	Kappaphycus alvarezii	Wheat	14.1% and 11.7% higher grain and straw yield respectively over 100% recommended dose of fertiliser (RDF)	Bhawesh *et al.* (2023)
8.	Ascophyllum nodosum	Strawberry	The seaweed extract in combination with humic acid significantly improved various growth parameters, including plant spread, number of leaves, and fruit yield. Increased level of total soluble solids, ascorbic acid, and total sugar content was observed	Chakraborty *et al.*, 2023

They had taken two types of N sources: (1) urea coated with a Nano polymer (Nano PCU) and (2) conventional urea. These were applied at three different rates (40,80, 120 kg ha^{-1}). From this study they found that the release pattern of inorganic N was significantly influenced by the polymer coating. The release pattern of Nano PCU in the laboratory displayed a gradual initial release followed by a rapid release (40-90 days) and ending with a slow rate of N release unlike conventional urea where rapid initial release (10 days) was followed by a sharp decline in release rate. Amanda *et al.* (2017) conducted two incubation studies to assess the influence of nanocomposites in regulating nitrogen availability in the soil. The hypothesis behind this was that nanocomposites can limit the liberation of urea in the soil and as a result act as a slow-release fertiliser. Nano composites were synthesised from thermoplastic starch/urea blend with hydroxyapatite particles (Hap). Nano composites resulted in a regulated liberation of urea and boosted the discharge of phosphorous from Hap. Maintenance of higher amount of mineral N due to application of slow-release brown coal urea fertiliser was reported by Saha *et al.* (2018). Li *et al.* (2021) reported that with the use of controlled release urea fertiliser, N accumulation in plants increased, thus improving agronomic efficiency and apparent recovery efficiency compared to conventional urea. Araujo *et al.* (2017) synthesised chitosan-based coating substances using peat, humin and humic acid to formulate a coated nitrogen fertiliser. Different humic substances influenced the discharge rate of N from the fertiliser composite due to the presence of various functional groups in the humic substances and their interaction with urea. Adlim *et al.* (2018) formulated a urea-magnesium-natural rubber composite plastered with chitosan. The study led to conclude that Mg intermingles with urea and gets confined in the rubber matrix, detaining urea release.

Precision Agriculture

It is the science of handling sophisticated, cutting-edge, novel site-specific technologies to administer spatial and temporal inconsistency in soil fertility to boost productivity, profitability and efficiency of agricultural land (Brar *et al.*, 2016). This technology differs from conventional uniform or blanket application methods, where fertilisers, pesticides and water are homogeneously spread across the land, frequently leading to inefficient use of agrochemicals (Raza *et al.*, 2023). It comprises various components such as remote sensing, global positioning system (GPS), geographic information system (GIS), data analytics, sensor technologies and machine learning. Variable rate technology enables us to reduce nutrient requirement, minimise waste generation, reduce cultivation costs, promote ideal crop growth and enhance environmental sustainability. Torrens-Sanchez *et al.* (2015) examined the application of drones in crop health monitoring and this study led to the conclusion that multi-spectral images captured by drones can precisely identify stressed vegetation. Sahu *et al.* (2017) examined the effect of nutrient management based on GIS-based soil fertility maps on soil fertility and wheat yield. Higher wheat yield and enhanced soil fertility were observed in the precision agriculture treated plots compared to the conventional uniform application rate, obviously due to site-specific input application giving due consideration to the variation in soil characteristics, nutrient levels, moisture content, topography and other factors. Zhang *et al.* (2019) conducted precision agriculture-based nutrient management in corn, where soil test based variable rate N and P fertilisers were applied considering the spatial variability of the field. Results revealed a significant increase in corn yield compared to the uniform blanket application method with considerable savings in agrochemicals. Fertiliser and pesticide application with the aid of unmanned aerial vehicles has been successfully reported from various parts of the world (Borikar *et al.*, 2022). With the help of machine-learning algorithms, predictive analysis about future crop yield forecasting and probable disease outbreaks can be done, enabling farmers to make quick decisions making by interpreting of the complex data (Luo *et al.*, 2020).

Besides these above-mentioned practices, several other innovative nutrient management practices have also been worked out by researchers all over the world. However, discussing of all of them is beyond the scope of this chapter. Scientists have dedicated themselves to discovering newer nutrient management practices, but the disappointing fact is that most of these attempts remain restricted in scientific publications only and the gap between lab and land still persists in a major portion of the world. Extension personnel have their roles to play here to bridge the gap between research laboratories and farmers' fields. Field trials and result demonstrations exhibiting the real practical advantages of these novel practices over the age-old traditional methods may prove effective in convincing the agrarians. Awarding and incentivising the obedient farmers might also be a good plan to consider.

Conclusion

To provide quality food for the increasing global population and at the same time maintain environmental sustainability, acknowledging the newer and smarter resource- conserving technologies is a must. Technologies like nano fertiliser formulations, biofertilisers, seaweed-based multi nutrient organic fertiliser, slow-release coated fertilisers, precision agriculture etc. no doubt have vast potential in alleviating global hunger, malnutrition and at the same time, can curtail the global ecological footprint by promoting efficient resource use. But the lack of expertise and technical know-how among farmers, and their apathy towards the newer innovations still remain major obstacles in their universalisation. Government schemes providing basic farmer training programmes, establishing custom hiring centres, incentivising obedient farmers and setting provisions for penalties for non-compliance might result in growing inquisitiveness among farmers about these practices.

References

Adhikary S, Mandal N, Rakshit R, Das A, Kumar V, Kumari N, Choudhary SK, Homa F. (2020) Field evaluation of zincated nanoclay polymer composite (ZNCPC): impact on DTPA-extractable Zn, sequential Zn fractions and apparent Zn recovery under rice rhizosphere. Soil and Tillage Research 201, 104607.

Adlim M, Zarlaida F, Rahmayani RFI, Wardani R. (2019) Nutrient release properties of a urea–magnesium–natural rubber composite coated with chitosan. Environment Technology & Innovation 16, 100442.

Afzal A, Bano A (2008) Rhizobium and phosphate solubilising bacteria improve the yield and phosphorus uptake in wheat (Triticum aestivum L.). International Journal of Agricultural Biology 10, 85– 88.

Amanda S, Giroto, MF, Ribeiro C. (2017) Role of slow-release nano composite fertilisers on nitrogen and phosphate availability in Soil. Scientific Reports 7, 46032.

Anantharaman P, Karthikaidevi G, Manivannan K, Thirumaran G, Balasubramanian T (2010) Mineral composition of marine macroalgae from mandapam coastal regions-southeast coast of India. Recent Research in Science and Technology 2, 66–71.

Araújo BR, Romão LP, Doumer ME, Mangrich AS (2017) Evaluation of the interactions between chitosan and humics in media for the controlled release of nitrogen fertiliser . Journal of Environmental Management 190, 122–131.

Arioli T, Mattner SW, Winberg PC. (2015) Applications of seaweed extracts in Australian agriculture: past, present and future. Journal of Applied Phycology 27, 2007–2015.

Bardi L, Malusà E. (2012) Drought and nutritional stresses in plant: alleviating role of rhizospheric microorganisms. abiotic stress: new research. Nova Science Publishers Inc., Hauppauge, pp. 1–57

Bardin SD, Huang HC, Pinto J, Amundsen EJ, Erickson RS. (2004). Biological control of Pythium damping-off of pea and sugar beet by Rhizobium leguminosarum bv. viceae. Can J Bot 82(3):291–296 Bashan Y (1998) Inoculants of plant growth promoting bacteria for use in agriculture. Biotechnology Advances 16(4), 729–770

Battacharya D, Babbohari MZ, Rathor P, Prithiviraj B. (2015). Seaweed extracts as biostimulants in horticulture. Scientia Horticulturae 196, 39–48.

Bernard B, Lux A. (2017). How to feed the world sustainably: An overview of the discourse on agroecology and sustainable intensification. Regional Environmental Change 17, 1279–1290.

Bhattacharyya PN, Jha DK. (2012). Plant growth-promoting rhizobacteria (PGPR): emergence in agriculture. World Journal of Microbiology and Biotechnology 28(4), 1327–1350

Bhawesh, B., Kumar, B., Mandal, N., Ghosh, M., & Sharma, A. (2023). Seaweed growth promoter can boost up the crop growth and enhance the economic yield of irrigated wheat. Journal of Soil and Water Conservation 22(1), 89-94.

Borikar, G. P., Gharat, C., & Deshmukh, S. R. (2022, October). Application of Drone Systems for Spraying Pesticides in Advanced Agriculture: A Review. In IOP Conference Series: Materials Science and Engineering (Vol. 1259, No. 1, p. 012015). IOP Publishing.

Brar, B. S., & Kaur, A. (2016). Precision Nutrient Management: A Review. Indian Journal of Fertiliser s 1, 15.

Chakraborty B, Basak S, Sherpa ZW, Samanta D, Apoorva ND, Gurung S. (2023). Effect Of Foliar Application of Humic Acid And Seaweed Extract In Strawberry (Fragaria x ananassa Duch.). Plant Archives 23(2).

Cole, J.C.; Smith, M.W.; Penn, C.J.; Cheary, B.S.; Conaghan, K.J. (2016). Nitrogen, phosphorus, calcium and magnesium applied individually or as a slow release or controlled release fertiliser increase growth and yield and affect macronutrient and micronutrient concentration and content of field-grown tomato plants. Scientia Horticulturae 211, 420–430.

Congreves KA, Otchere O, Ferland D, Farzadfar S, Williams S, Arcand MM (2021). Nitrogen use efficiency definitions of today and tomorrow. Frontiers of Plant Science 12, 637180

Deshwal VK, Pandey P, Kang SC, Maheshwari DK. (2003). Rhizobia as a biological control agent against soil borne plant pathogenic fungi. Indian J Exp Biol 41(10), 1160–1164.

Dhargalkar VK, Pereira N. (2005). Seaweed: promising plant of the millennium. Scientific Culture 71, 60–66.

Dhargalkar VK, Untawale AG, Jagtap TG. (2001) Marine macroalgal diversity along the Maharashtra coast: past and present status. Indian Journal of Marine Sciences 30, 18–24

Dimkpa, C. O., Singh, U., Adisa, I. O., Bindraban, P. S., Elmer, W. H., Gardea-Torresdey, J. L., & White, J. C. (2018). Effects of manganese nanoparticle exposure on nutrient acquisition in wheat (Triticum aestivum L.). Agronomy 8(9), 158.

Dogra BS, Mandradia RK. (2012) Effect of seaweed extract on growth and yield of onion. International Journal of Farm Sciences 2(1):59–64

El-Miniawy, S. M., Ragab, M. E., Youssef, S. M., & Metwally, A. A. (2014). Influence of foliar spraying of seaweed extract on growth, yield and quality of strawberry plants. Journal of Applied Sciences Research 10(2), 88-94.

Elsayed AAA, Ahmed EG, Taha ZK, Farag HM, Hussein MS, AbouAitah K (2022). Hydroxyapatite nanoparticles as novel nano-fertiliser for production of rosemary splants. Scientia Horticulturae 295, 110851.

Fageria N K, Baligar V C, Li YC. (2014). Nutrient uptake and use efficiency by tropical legume cover crops at varying pH of an Oxisol. Journal of Plant Nutrition 37, 294-311.

Giller, K.E., Delaune, T., Silva, J.V., Descheemaeker, K., van de Ven, G., Schut, A.G.T., van Wijk, M., Hammond, J., Hochman, Z., Taulya, G., Chikowo, R., Narayanan, S., Kishore, A., Bresciani, F., Teixeira, H.M., Andersson, J.A., van Ittersum, M.K. (2021). The future of farming: who will produce our food, Food Security. 13, 1073–1099.

Gil-Ortiz, R.; Naranjo, M.Á.; Ruiz-Navarro, A.; Atares, S.; García, C.; Zotarelli, L.; San Bautista, A.; Vicente, O (2020). Enhanced Agronomic Efficiency Using a New Controlled-Released, Polymeric-Coated Nitrogen Fertiliser in Rice. Plants 9, 1183

Khan AL, Waqas M, Kang SM. (2014) Bacterial endophytes Sphingomonas sp LK11 produces gibberellins and IAA and promotes tomato plant growth. Journal of Microbiology 52, 689–695.

Khan W, Rayirath UP, Subramanian S, Jithesh MN, Rayorath P, Hodges DM, Critchley AT, Craigie JS, Norrie J, Prithiviraj B. (2009) Seaweed extracts as biostimulants of plant growth and development. Journal of Plant Growth Regulation 28, 386–39.

Khodakovskaya M, Dervishi E, Mahmood M, Xu Y, Li Z, Watanbe F, Biris AS (2009) Carbon nanotubes are able to penetrate plant seed coat and dramatically affect seed germination and plant growth. ACS Nano 3, 3221–3227.

Khodakovskaya M, Silva KD, Biris AS, Dervishi E, Villagarica H (2012) Carbon nanotubes induce growth enhancement in tobacco cells. ACS Nano 6, 2128–2135.

Kloepper JW, Leong J, Teintze M, Schroth MN (1980) Enhanced plant growth by siderophores produced by plant growth-promoting rhizobacteria. Nature 286(5776), 885–886

Kurepa J, Paunesku T, Vogt S, Arora H, Rabatic BM, Lu J, Wanzer MB, Woloschak GE, Smalle JA. (2010) Uptake and distribution of ultrasmall anatase TiO2 Alizarin red S nanoconjugates in Arabidopsis thaliana. Nano Lett 10, 2296–2302.

Li Z, Liu Z, Zhang M, Chen Q, Zheng L, Li YC, Sun L. (2021) The combined application of controlled-release urea and fulvic acid improved the soil nutrient supply and maize yield. Archives of Agronomy and Soil Science 67(5), 633-646.

Lu CM, Zhang CY, Wu JQ, Tao MX (2002) Research of the effect of nanometre on germination and growth enhancement of Glycine max and its mechanism. Soybean Science 21, 168–172

Luo, J., Wang, Y., & Zou, J. (2020). Big data in precision agriculture. Journal of Integrative Agriculture 19(11), 2605-2616.

Mahanty, T., Bhattacharjee, S., Goswami, M. *et al.* (2017). Biofertilisers: a potential approach for sustainable agriculture development. Environmental Science and Pollution Research 24, 3315–3335.

Malusa E, Vassilev N. (2014) A contribution to set a legal framework for biofertilisers. Applied Microbiology and Biotechnology 98(15), 6599–6607.

Mandal N, Datta SC, Manjaiah KM, Dwivedi BS, Kumar R, Aggarwal P (2018) Zincated nanoclay polymer composites (ZNCPCs): synthesis, characterisation, biodegradation and controlled release behaviour in soil. Polymer-Plastics Technology and Material. Eng 57, 1760–1770.

Mazid M, Khan TA. (2015) Future of bio-fertiliser s in Indian agriculture: an overview. International Journal of Agricultural and Food Research 3(3), 10–23.

Milton RF (1952) Improvements in or relating to horticultural and agricultural fertiliser s. Br Patent 664, 989.

Mohy El-Din SM. (2015) Utilisation of seaweed extracts as bio-fertilisers to stimulate the growth of wheat seedlings. Egyptian Journal of Experimental Biology 11(1), 31–39.

Mondal A, Basu R, Das S, Nandy P. (2011) Beneficial role of carbon nanotubes on mustard plant growth: an agricultural prospect. Journal of Nanoparticle Research 13, 4519–4528

Nabti E, Sahnoune M, Ghoul M, Fischer D, Hofmann A, Rothballer M, Schmid M, Hartmann M. (2010) Restoration of growth of durum wheat (*Triticum durum* var. *waha*) under saline conditions due to inoculation with the rhizosphere bacterium Azospirillum brasilense NH and extracts of the marine alga Ulva lactuca. Journal of Plant Growth Regulation 29, 6–22.

Nair R, Varghese SH, Nair BG, Maekawa T, Yoshida Y, Kumar DS. (2010) Nanoparticulate material delivery to plants. Plant Science 179, 154–163.

Nowak TB, Gould SJ, Kraus J, Loper JE. (1994) Production of 2,4- diacetyl phloroglucinol by the biocontrol agent Pseudomonas fluorescens. Canadian Journal of Microbiology 40(12), 1064–1066

Osman SA, Salama DM, Abd El-Aziz ME, Shaaban EA, Abd Elwahed MS. (2020). The influence of MoO3-NPs on agro-morphological criteria, genomic stability of DNA, biochemical assay, and production of common dry bean (Phaseolus vulgaris L.). Plant Physiology and Biochemistry. 151, 77–87.

Patil MD, Das BS, Barak E, Bhadoria PBS, Polak A (2010) Performance of polymer-coated urea in transplanted rice: effect of mixing ratio and water input on nitrogen use efficiency. Paddy Water Environment, 8, 189–198.

Prabha C, Maheshwari DK, Bajpai VK. (2013). Diverse role of fast growing Rhizobia in growth promotion and enhancement of psoralen content in Psoralea Corylifolia L. Pharmacogn Mag 9(Suppl 1):S57.

Raghunandan, B.L., Vyas, R.V., Patel, H.K., Jhala, Y.K. (2019). Perspectives of Seaweed as Organic Fertiliser in Agriculture. In: Panpatte, D., Jhala, Y. (eds) Soil Fertility Management for Sustainable Development. Springer, Singapore.

Raza, I., Zubair, M., Zaib, M., Khalil, M. H., Haidar, A., Sikandar, A., ... & Ashfaq, M. A. (2023). Precision Nutrient Application Techniques to Improve Soil Fertility and Crop Yield: A Review with Future Prospect. International Research Journal of Education and Technology. 05 (08): 109-123.

Roberts TL, Johnston AE. (2015). Phosphorus use efficiency and management in agriculture. Resources, Conservation and Recycling 105, 275-281.

Rokana S, Mandal N, Singh M, Pradhan AK, Ghosh M, Sadhu S, Kumari S, Kumar R, Kumar V. (2023) Synthesis, Characterization and Controlled Release Pattern of Novel Fe Containing Nanoclay Polymer Composites (Fe NCPC) in Non calcareous and Calcareous Soil. Journal of Soil Science and Plant Nutrition, 1-16.

Rosalba Mireya Hernández-Herrera FS-R, Ruiz-López MA, Norrie J, Hernández-Carmona G. (2014) Effect of liquid seaweed extracts on growth of tomato seedlings (Solanum lycopersicum L.). J Appl Phys 26(1), 619–628.

Sabry SR, Saleh SA, Batchelor CA, Jones J, Jotham J, Webster G, Kothari SL, Davey MR, Cocking EC. (1997). Endophytic establishment of Azorhizobium caulinodans in wheat. Proceedings of the Royal Society of London B: Biological Sciences 264(1380), 341–346.

Sadhu S, Sow S, Ranjan S, Sahoo R. (2023). Polymer Coated Fertiliser s and their Scope in Modern Agriculture. In: Devi KM *et al* (eds) Current Trends and Advances in Agricultural Sciences. Kripa-Drishti Publications, pp 42-53.

Safinaz AF, Ragaa AH. (2013) Effect of some red marine algae as biofertilisers on growth of maize (Zea mayz L.) plants. International Food Research Journal 20(4), 1629–1632.

Saha BK, Rose MT, Wong VN, Cavagnaro TR, Patti AF (2018) Nitrogen dynamics in soil fertilized with slow-release brown coal-urea fertilisers. Scientific reports 8(1), 1-10.

Sahu, A., Shukla, A. K., Rao, A. S., Ramesh, V., Gupta, P. K., & Meena, R. S. (2017). Precision nutrient management in wheat for enhancing productivity, nutrient-use-efficiency, and soil fertility. Communications in Soil Science and Plant Analysis 48(8), 970-983.

Sarkar, S., Datta, S.C. & Biswas, D.R. (2015). Effect of Fertiliser Loaded Nanoclay/ Superabsorbent Polymer Composites on Nitrogen and Phosphorus Release in Soil. Proceedings of the National Academy of Sciences, India, Sect. B Biol. Sci. 85, 415–421.

Shaharoona B, Arshad M, Zahir ZA (2006) Effect of plant growth promoting rhizobacteria containing ACC-deaminase on maize (*Zea mays* L.) growth under axenic conditions and on nodulation in mung bean (*Vigna radiata* L.). Letters in Applied Microbiology. 42(2), 155–159.

Shevchenko NM, Anastyuk SD, Gerasimenko NI, Dmitrenok PS, Isakov VV, Zvyagintseva TN. (2007) Polysaccharide and lipid composition of the brown seaweed Laminaria gurjanovae. Russian Journal of Bioorganic Chemistry. 33, 88–98.

Simonet P, Normand P, Moiroud A (1990) Identification of Frankia strains in nodules by hybridization of polymerase chain reaction products with strain-specific oligonucleotide probes. Archives of Microbiology. 153(3), 235–240.

Singh T, Kumar Y, Sain NK, Laxmi S. (2022) Innovative Specialty and Nano Fertiliser s for Enhancing Nutrient Use Efficiency in the Dryland Agro-ecosystems. Indian journal of fertilisers. 18 (11), 1131-1144.

Sivasangari Ramya S, Vijayanand N, Rathinavel S. (2015) Foliar application of liquid biofertiliser of brown alga Stoehospermum marginatum on growth, biochemical and yield of Solanum melongena. International Journal of Recycling of Organic Waste in Agriculture. 4(3), 167–173

Ţolescu, C., Iovu H. (2010). Polymer Conditioned Fertilisers", U.P.B. Sci. Bull., Series B 72 909

Torres-Sánchez, J., López-Granados, F., de Castro, A. I., & Peña-Barragán, J. M. (2015). Configuration and specifications of an unmanned aerial vehicle (UAV) for early site specific weed management. PLoS ONE, 10(3), e0117659.

Tuhy Ł, Samoraj M, Basadynska S, Chojnacka K. (2015) New micronutrient fertiliser biocomponents based on seaweed biomass. Polish Journal of Environmental Studies. 24, 2213–2

Upadhyay PK, Dey A, Singh VK, Dwivedi BS, Singh T, G. A. R, *et al.* (2023) Conjoint application of nano-urea with conventional fertilisers: An energy efficient and environmentally robust approach for sustainable crop production. PLoS ONE 18, e0284009.

Uzu G, Sobanska S, Sarret G, Munoz M, Dumat C. (2010) Foliar lead uptake by lettuce exposed to atmospheric fallouts. Environmental Science and Technology. 44, 1036–1042.

Verma JP, Yadav J, Tiwari KN, Lavakush SV (2010) Impact of plant growth promoting rhizobacteria on crop production. International Journal of Agricultural Research. 5, 954–983.

Wang, Y.; Liu, M.; Ni, B.; Xie, L. (2012) κ-Carrageenan–sodium alginate beads and superabsorbent coated nitrogen fertiliser with slow-release, water-retention, and anti-compaction properties. Industrial & Engineering Chemistry Research 51, 1413–1422.

Yang X, Geng J, Li C, Zhang M, Tian X. (2016) Cumulative release characteristics of controlled-release nitrogen and potassium fertilisers and their effects on soil fertility, and cotton growth. Scientific Reports 6(3), 62- 75.

Zhang, Y., Zhu, X., Luo, Z., Zhang, J., Chen, S., Yang, S., & Cao, W. (2019). Effects of site-specific nutrient management on nutrient use efficiency, soil fertility, and corn yield. Journal of Integrative Agriculture 18(1), 178-189.

15

Regenerative Practices for Restoring the Soil Fertility

Sumit Kumar[1], Rishikesh Yadav[1]
Praveen Kumar Yadav[2] and Sone Lal Patel[3]

[1]*Department of Soil Science and Agricultural Chemistry, Chandra Shekhar Azad University of Agriculture and Technology, Kanpur 208002, Uttar Pradesh*

[2]*Department of Agronomy, Chandra Shekhar Azad University of Agriculture and Technology, Kanpur 208002, Uttar Pradesh*

[3]*Department of Agronomy, Chimanbhai Patel College of Agriculture, Sardarkrushinagar Dantiwada Agricultural University, S K Nagar 385506, Gujarat*

Abstract

Regenerative agriculture is described as holistic farming systems that, among other benefits, improve water and air quality, enhance ecosystem biodiversity, produce nutrient-dense food, and store carbon to help mitigate the effects of climate change. Restoring soil fertility is a critical imperative for sustainable agriculture, and explores regenerative practices as a comprehensive solution. The understanding of soil health, emphasising microbial life, soil structure, and nutrient balance, establishes a foundation for holistic approaches. Cover cropping and crop rotation are discussed, showcasing their roles in protecting soil, enhancing biodiversity, and breaking pest cycles for improved nutrient availability. The principles of no-till farming emerge as a key contributor to soil preservation, reducing disturbance and enhancing fertility. Regenerative grazing is examined for its impact on soil structure and nutrient cycling, emphasising sustainable livestock management practices. Biochar and soil amendments are investigated for their potential in enhancing fertility and sequestering carbon. Community engagement is emphasised as a pivotal aspect, showcasing the significance of collaborative knowledge-sharing initiatives. Cover cropping and crop rotation take center stage, offering multifaceted benefits such as erosion prevention, biodiversity enhancement, and improved nutrient availability. Agro-forestry and alley cropping systems are examined for their ability to augment biodiversity, control erosion, and contribute holistically to soil fertility. Using regenerative agriculture practices not only assists farmers in mitigating the effects of present climate change, but also enables them to take long-term action by participating in a wider solution to the crisis through carbon sequestration. It also makes their farms more resilient and adaptive to what is happening around them now.

Keywords: *Biochar, Biodiversity, Climate change, Carbon sequestration, Regenerative Agriculture, Soil Health*

Introduction

Soil fertility is a key factor in successful crop production measures the ability of the soil to supply nutrients to plants. Soil fertility and fertilisers are closely related concepts. Soil fertility acts as a "SOURCE" from which plants can absorb nutrients for maximum yield, while manure acts as a "SOURCE" from which we can continuously absorb various nutrients and also add to the soil. The importance of soil fertility and fertiliser management has been increasingly recognised in all countries to meet the demand for food and other agricultural commodities. Intensive use of fertilisers and cultivation of high-yielding varieties has undoubtedly increased food production and reduced food scarcity, but it has also brought with it a host of problems related to soil fertility, soil and water pollution (Sow *et al.*, 2023a). On the other hand, rapid depletion of nutrients due to overuse, widespread N, P, K, and S deficiency, along with micronutrient deficiencies, especially Zn and Boron, have been observed in many soils. The implications of soil fertility extend across various aspects of the Sustainable Development Goals, encompassing economic, social, and environmental dimensions. A fertile soil plays a crucial role in addressing the Zero Hunger objective set by the FAO, primarily by providing the essential function of producing food. This food, cultivated in fertile soil, contains the necessary nutrients vital for human health. Beyond its contribution to food production, soil fertility plays a pivotal role in activities with economic implications, influencing economic growth and contributing to poverty alleviation. Furthermore, effective management of soil fertility holds the potential to mitigate soil, water, and air pollution, regulate water resources availability, support a diverse and active biotic community, enhance vegetation cover, and promote a carbon-neutral footprint. In regions like India, experiencing a rapid decline in the land-to-population ratio, achieving increased productivity without compromising sustainability and environmental integrity becomes imperative (Ranjan *et al.*, 2023a; Sow and Kumar, 2023). This approach is essential to meet the growing demand for agricultural output in emerging nations and underscores the interconnectedness of soil fertility with broader developmental goals. From 0.48 ha in 1951 to 0.14 ha in 2001, the country's per capita land availability has declined, and by 2025, it is expected to have further decreased to 0.10 ha. Due to urbanisation, industrialisation, and other factors, there is less land available for cultivation overall, as well as less high-quality land suitable for crops. This has resulted in the cultivation of marginal lands and the removal of forests, impoverishing the quality of the soil and increasing the susceptibility of the land to erosion and floods. As a result, the country and its productive resource base are adversely affected. This requires regulating soil fertility along with reducing soil quantity and quality.

Soil fertility refers to the ability of the soil to provide plants with the nutrients they need for optimal growth in sufficient quantities, in the right proportions, and at the right time. Managing soil fertility requires building and maintaining it at a high level to produce enough food, fodder, and fuel for an ever-increasing human and animal population (Sadhu *et al.*, 2023). Therefore, fertilisers are of great importance and one of the most important factors to achieve high productivity. India does not produce enough primary plant nutrients on its own. A significant portion of phosphatic and nitrogenous fertilisers, as well as the whole amount of potassic fertilisers, are imported. Scientists are therefore more concerned about the efficient management of soil fertility due to the growing environmental danger and the growing gap between fertiliser production and use (Sow *et al.*, 2023b). Knowing the nutritional requirements of plants, how nutrients cycle through soils, the limitations on soil fertility, and effective nutrient management are all necessary for effective soil fertility management (Kumar *et al.*, 2023).

Current Challenges in Soil Health

India is a fast-developing economic system with competencies to make contributions to global food security and meet sustainable improvement goals (SDGs). However, inclusive financial boom in India faces the demanding situations of balancing among city and rural regions amid the daunting undertaking of feeding an ever-growing population. The presence of organic matter in soils, particularly the stock of soil organic carbon (SOC), is crucial for maintaining soil health and supporting ecosystem functions. Carbon is also a very important element for soil structure and fertility. Carbon and nitrogen are the basic elements that provide a food source for soil life. Soils are home to an estimated 25% of the world's biodiversity and merely a teaspoon of healthy soil contains about one billion microorganisms. Most people in India are still dependent on agriculture and livestock for their lives and livelihoods. Healthy soil is not only important for producing healthy food but also for combating the adverse impacts of climate change. The combination of a high population, high agricultural production and diverse agro-climatic conditions creates more pressure on land ,resulting in an increase in risks related to land degradation. Humans worldwide obtain more than 99.7% of their food from the soil. Further the ever-deterioration of soil health across the globe has reduced the productivity of all natural ecosystems such as water availability, agriculture, forest and pasture ecosystems. The reduced productivity of ecosystems has also affected the nutrition levels of human beings. According to an estimate today around 66% of the world's population is malnourished. Therefore, for the well-being of humans, wildlife and various other ecosystems keeping good health of soil is important. The term soil health refers to properties of soil and the critical functions of the soil. The Indian Council of Agricultural Research and the National Academy of Agricultural Sciences in 2010 published a detailed data set on soil erosion and the major factors affecting soil health. The data set reveals that nearly 86 mha of the arable land is deteriorating due to water and wind erosion. Further more than 17 mha of arable land is deteriorating due to chemical degradation. In total 120.72 mha of arable land and open forest is under soil erosion due to various factors.

Regenerative agriculture prioritises soil over seeds and is founded on the principle that the "health of soil, plants, animals, and humans is one and indivisible" (Howard, 1943; Howard and Howard 1945). In the current context of climate change and environmental concerns, it is pertinent to broaden the concept to emphasise that the health of soil, plants, animals, people, and the environment is interconnected. This expanded notion, grounded in the understanding of the living soil (Balfour, 1943), holds particular relevance in the era of COVID-19 (Lal, 2020). Coleman and Crossley (1996) highlight that "one of the last great frontiers in biological and ecological research is the soil." Shaxson (1997) contends that the challenge confronting scientists, policymakers, and farmers worldwide is to find ways to increase food production on diminishing land, with fewer opportunities for land recuperation fallow periods, a net loss of nutrients, and a rise in crop pests and plant pathogen populations. Land productivity will increasingly rely on soil amendments and intensive management measures to control pest populations. Rooted in biological principles, regenerative agriculture aims to simultaneously enhance productivity and environmental management (Pretty, 1995).

Regenerative Agriculture

The term 'regenerative' has been linked with the nouns 'agriculture' and 'farming' since the late 1970s (Gabel, 1979). However, the terms Regenerative Agriculture and Regenerative Farming gained wider circulation in the early 1980s when the US-based Rodale Institute adopted them. Over the years, through its research and publications, including the magazine Organic Gardening and Farming, the Rodale Institute has played a leading role in the organic farming movement. Rodale (1983) defined Regenerative Agriculture as "one that, at increasing levels of productivity, increases our land and soil biological production base. It has a high level of built-in economic and biological stability. It has minimal to no impact on the environment beyond the farm or field boundaries. It produces foodstuffs free from biocides. It provides for the productive contribution of increasingly large numbers of people during a transition to minimal reliance on non-renewable resources."

"Regenerative Agriculture" denotes farming and grazing practices that, among other benefits, reverse climate change by rebuilding soil organic matter and restoring degraded soil biodiversity, resulting in both carbon drawdown and improvements in the water cycle. Specifically, regenerative agriculture is a holistic land management practice that uses the photosynthetic power of plants to close the carbon cycle and build soil health, crop tolerance and nutrient density. Regenerative agriculture improves soil health, mainly through practices that increase soil organic matter (Khangura *et al.*, 2023). Not only does it help increase the diversity and health of soil biota, but it also increases the biodiversity both above and below ground level increasing both water retention and binding carbon dioxide at greater depths, which reduces atmospheric levels harmful to the climate CO2 and improves soil structure to reverse anthropogenic soil-threatening . Research continues to investigate the harmful effects of cultivation and application of agricultural chemicals and salt-based fertilisers and coal mining. Regenerative Agriculture transforms this paradigm to build the future (Giller *et al.*, 2021). This is a considered solution for many of the most pressing issues, including soil and water pollution, food shortages and climate change (Sahoo *et al.*, 2023). Its characteristic is good soil restoration that leads to healthy land, healthy animals, crop resistance and nutrient density and ultimately healthier food for all. It not only helps increase the diversity and health of the soil biota, but also increase biodiversity both upstream and downstream on the surface of the soil and add water at the same time storage capacity and carbon footprint greater depths, which reduces climate-damaging effects atmospheric CO2 levels and soil improvement the structure overturns the structure that threatens civilisation anthropogenic soil loss.

Principles of Regenerative Agriculture

Work with whole systems, not isolated parts and connect the farm to its larger ecosystem and bio-region. It is a system that aims to imitate nature, not constantly trying to beat it. By reversing the cycle of soil degradation, regenerative agriculture can actually replenish the soil carbon concentrations. It seems like the word "regenerative" is becoming a buzzword for sustainable corporate development but "regenerative agriculture" is a distinct movement in agriculture. The five key principles of regenerative agriculture encompass all areas on the farm, and within each of these are the practices that can be adopted to transition towards more sustainable agriculture. The five principles are:

1. Minimise soil disturbance: Since the beginning of time, farmers have believed that tillage or soil loosening, would improve soil fertility, increase the soil's capacity to absorb rainwater, and aid in the control of undesired weed flora (Roy *et al.*, 2023a). This has led to the widespread practice of plowing fields to prepare seed beds or for sowing. Because of the physical, chemical, and biological deterioration of the soil, this practice eventually results in a decrease in soil fertility and productivity. Repeated tillage operations erode soil properties, leaving it more susceptible to erosion and runoff. Organic matter is essential for maintaining a stable soil structure. So, minimal soil disturbance is encouraged by regenerative agriculture, which means that very little or no tilling is done and also try to reduce or eliminate the use of chemicals such as pesticides and chemical fertilisers.
2. Maximise crop diversity: Diversity helps build healthy soil that better binds water and nutrients, can provide other sources of income for the farm, and can benefit pollinators and wildlife. Regenerative farms can vary crop rotation, plant several cover plant species together, grow versatile forage in the pasture and maintain permanent vegetation (protective cover) in some farm zones.
3. Keep the soil covered: Instead of cultivating the soil, regenerative practices include mulching, planting cover crops (plants that are not sold but provide other benefits such as soil improvement, water retention, weed control, and erosion prevention), and maintaining the land as permanent pasture (Roy *et al.*, 2023b).
4. Maintain living root year round: Roots stabilise the soil and constantly redirect water and nutrients to prevent these valuable resources from being washed away. Regenerative farms can achieve this by planting cover crop seeds in the same fields as the main crops before harvest; so that the fields are never bare (a method called overseeding). Planting primary crops directly in fields where cover crops are already growing (so-called "green" planting as cover crops); or convert cropland to pasture.
5. Integrate livestock: Animal manure can add valuable nutrients to the soil, reducing the need for fertilisers, and permanent pastures can sequester large amounts of carbon and water, reducing farm emissions and polluted runoff. Practices include rotational grazing often moving cattle between grass pastures to give plants time to recover or cover crop grazing.

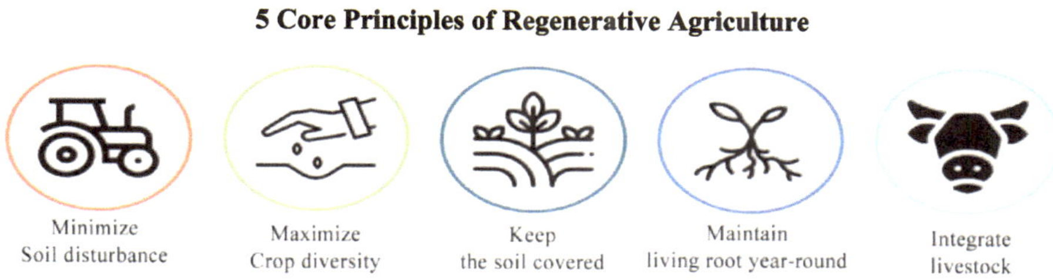

Fig. 1: Principles of Regenerative agriculture

Need for Regenerative Agriculture Practices

Regenerative agriculture practices bridge the gap between traditional, organic and conventional farming practices. With various pressures on cultivating crops, whether it be the rising fertiliser prices, threats of climate change or the more extreme weather, farming practices are transitioning towards less conventional methods (Schreefel *et al.*, 2022). Today's agriculture contributes to soil loss and deterioration through the use of pesticides, fertilisers, and large machinery to increase food production. According to Regeneration International, an organisation that promotes regenerative farming, there may not be enough soil left in 50 years to feed everyone on the planet. Additionally, CO_2 that is naturally held in soil is disturbed and released into the atmosphere by intensive cultivation. This adds to the warming of the planet that is causing climate change (Sow and Ranjan, 2021). The United Nations (UN) estimates that more than one-third of greenhouse gas emissions worldwide are caused by agriculture. On the other hand, as the Earth heats, extreme weather events like flooding are becoming more frequent and intense, making habitats more susceptible to damaged soil and eroding land. Embracing a restorative and regenerative approach to agricultural production is imperative for ensuring the economic and environmental resilience of future food systems. Merely aiming for 'sustainability' falls short, especially when 38% of global agricultural land is already affected by degradation. The depletion of fertile soil and biodiversity, coupled with the loss of indigenous seeds and knowledge, poses a severe threat to our future survival. Soil scientists warn that at the current rates of soil destruction—through decarbonisation, erosion, desertification, and chemical pollution—within 50 years, we will face not only significant damage to public health due to a qualitatively degraded food supply with diminished nutrition and loss of crucial trace minerals but also a literal shortage of arable topsoil to feed the global population.

Preserving and regenerating the soil on our 4 billion acres of cultivated farmland, 8 billion acres of pastureland, and 10 billion acres of forest land is paramount. This is essential not only for feeding the world but also for mitigating global warming, keeping it below 2 degrees Celsius, and halting the loss of biodiversity. Farmers, grappling with unpredictable weather patterns, rising costs of fertilisers and energy, and the need for more sustainable operations, find in regenerative agriculture an ally. Beyond environmental benefits, regenerative practices contribute to increased farm profitability, reduced input costs, heightened resilience, and job creation in rural areas. Regenerative agriculture stands out as a holistic solution to restore and enhance soil health, addressing the fundamental dependence of agricultural ecosystems on soil fertility (Ranjan and Sow, 2021). In contrast to conventional farming practices that often deplete soils, regenerative approaches hold the promise of sustainable agriculture while fostering economic opportunities and prosperity in rural communities.

Regenerative Agricultural Practices

In general, adopting regenerative practices means-"Same Land, Different Management".Regenerative agricultural practices focus on working with nature, limiting expensive artificial inputs, and mimicking natural ecosystems in an agricultural environment. It bases its practices on agro-ecology, permaculture and conservation agriculture; its purpose is to restore soil health. Some traditional farmers have joined innovative agricultural practices such as reduced tillage, crop rotation and cover cropping for years. By adopting innovative agricultural practices, soils also often show an increase in carbon compared to nearby facilities. When special techniques are used in well-structured agriculture the order reduces the farmer's dependence on external inputs, and management is less determined by external influences that the farm family has little control over. A farmer can implement these practices by partially replacing this knowledge or with new information about what has been purchased in the past to increase performance.

Table 1. Selected regenerative agricultural systems and how they include five regenerative agriculture practices

System	Minimise tillage	Minimise bare soil	Foster plant diversity	Integrate crops and animals	Synthetic fertilizers/ pesticides
Conservation agriculture	√	√			
Organic crop production	√	√	√		x
Tree crops	√	√			
Tree intercropping	√	√	√		
Multistrata agro-forestry	√	√	√		

Silvopasture	√	√	√	√	
Multi-paddock grazing	√	√	√	√	
Organic grassland systems	√	√	√	√	x
Rewilding	√	√			

Legend: √ means necessary; x means prohibited; blank space means optional

There are many regenerative practices that fulfil a regenerative philosophy:

1. No-till farming/minimal tillage: No-till farming has been in existence since the 1960s, when it was introduced to aid in halting the loss of farmland's valuable topsoil. Fewer disruptions to the soil allow more diverse soil microbes that provide better soil structure for your plants to grow. Tillage breaks up (pulverizes) soil aggregation and fungal groups while including extra O_2 to the soil for increased respiratory and CO_2 emission. It may be one of the most degrading agricultural practices, substantially increasing soil erosion and carbon loss. A secondary impact is soil capping and slaking that may plug soil areas for percolation growing a good deal extra water runoff and soil loss. Conversely, no-till/minimal tillage, in conjunction with different regenerative practices, complements soil aggregation, water infiltration and retention, and carbon sequestration (Dayal *et al.*, 2023). However, a few soils gain from intervening time ripping to break aside hardpans, that can growth root zones and yields and feature the potential to growth soil fitness and carbon sequestration. Certain low-level chiseling may also have similar beneficial effects.
2. Crop rotation and cover crops: Crop rotation, also known as diversification, is a centuries old practice which improves yield and profit by providing nutritional benefits and breaking the pest–disease–weed cycle. If soil is left exposed to the elements, it will erode substances and nutrients required for the plant growth either dries up or washes away. The same plants are planted at the same time the same place can lead to excess some nutrients and lack of others. Farms and gardens can replenish the soil with more (and more diverse) soil organic matter through crop rotation and careful propagation of cover crops, often naturally preventing disease and reducing pest concerns. Crop rotation moderates soil nutrients through growing diverse crops with different nutrient inputs and requirements (Stockdale *et al.* 2001). Always remember that bare soil is unhealthy. Artificial and synthetic fertilisers disturb the structure and function of soil microbial communities, bypassing the natural biological consumption of nutrients by plants,

resulting in a dependent agro ecosystem and weaker, less flexible plants. Cover crops, crop rotation, compost and manure restore the plant/ soil micro biome, which supports the release, transfer and circulation of the most important nutrients in the soil, increasing soil fertility in regenerative systems physiologically (Shao *et al.*, 2023).

3. Well-managed grazing practices: Another popular regenerative practice is to integrate livestock to improve soil health and diversify the income stream, despite the fact that livestock farming is widely blamed for contributing to methane emissions. Opting for rotational grazing is the preferred method over continuous grazing to enhance soil health and increase soil organic carbon (SOC). These practices increase soil fertility, insect and plant biodiversity, and soil carbon sequestration while also stimulating improved plant development, higher soil carbon deposits, and overall pasture and grazing area productivity (Ranjan *et al.*, 2023b). These practices benefit not only the environment, but also the animal and human consumers' health by increasing micronutrient availability and improving dietary omega balances. Feed lots and confined animal feeding systems significantly contribute to (i) unhealthy monoculture production systems, (ii) low nutrient density fodder, (iii) increased water pollution, (iv) antibiotic use and resistance, and (v) CO2 and methane emissions, all of which combine to produce broken and ecosystem degrading food production systems. Organic livestock are raised in a different way, with a focus on rotational grazing, which involves rotating animals through pastures to promote the health of soil, plants, and animals. Animal welfare is of the foremost importance, with the goal of meeting the animals' needs and maintaining health without the use of antibiotics or hormones.
4. Composting and Compost Tea: Composting is the natural process of turning waste (from manure or food) into valuable manure. Restoring depleted soils requires establishing an organic soil structure. Among the various options for waste recycling, composting is widely followed for handling diverse organic materials. Compost can be produced from waste materials of various kinds, such as cereal straws, crop stubbles, cotton stalks, groundnut husk, farm weeds and grasses, leaves, leaf-mould, biodegradable segment of household, factories and food processing plants refuse, wood ash, litter, urine-soaked earth from cattle-sheds and other similar substances. These raw materials are rich in cellulose and other readily decomposable carbohydrates and have a C/N ratio of 40 or more to 1 and the most efficient composting occurs with an optimal C/N ratio of about 25:1. In regenerative agriculture, composted

biological materials such as crop residue, food waste, and animal waste are essential for boosting soil organic matter levels (Sow *et al.*, 2020). Creating biological ecosystem diversity begins with soil inoculation with composts or compost extracts to restore the soil microbial community population, structure and functionality and to restore soil energy (C-compounds such as exudates). This can be achieved by planting full-time between several crops vegetable plantations, multi-species cover crops, bee plantations habitats and other beneficial insects. This can include very successful push-pull systems.

The term "compost tea" refers to a wide range of preparations that start with compost and create a liquid extract of the original compost. Compost tea is likely made in a variety of ways, with an emphasis on its application in more specialised fields. A more accessible form of compost that will have an effect on the plant sooner than compost incorporated into the soil is compost tea. Compost teas are sprayed on the foliage of plants or the soil. Those that are sprayed on the soil will seep into the root zone and impact the plant's rhizosphere.

5. Agro-forestry and Alley cropping: The purposeful incorporation of trees and shrubs into crop and animal farming systems with the goal of generating economic, social, and environmental benefits is known as agro-forestry. Each of these various components works well with the others. An Indigenous practice involves mimicking forest systems by integrating trees and shrubs into crop and animal systems. Although agro-forestry includes rangeland, a wider range of tools is used to transform large agricultural landscapes in ways that have far-reaching environmental, social and economic impacts. Agroforestry also includes the cultivation of silvopasture between rows of trees, forest management, buffers between riparian forests, watersheds in fields and windbreaks. Alley cropping, an alternative to traditional farming systems is established by perennials in association with annuals. The annuals are grown between the hedgerows of planted trees and shrubs, preferably fast-growing, leguminous species established at 4 to 8 m apart, which are periodically pruned to prevent shading the companion crop(s). When these practices are combined in a purposefully designed and managed manner, they increase plant diversity, soil health, reduce agricultural runoff, protect against soil erosion, and provide habitat for native flora and fauna. These practices also enhance biodiversity, nutrient cycling and contribute organic matter (Nath *et al.*, 2023a). Economically, these practices provide additional income through the use of practices such as forest farming, growing second crops between tree rows, using wildlife

in riparian forest buffers, and allowing for the combination of grazing with trees.

The three types of agro-forestry are:

i) Agrisilvicultural systems- Agriculture crops are combined with trees

ii) Silvopastoral systems- Trees combined with either with pastures or animals

iii) Agrosilvipastoral systems- Combination of all three components (trees, crops and animals)

6. Biochar and Terra preta: Biochar and terra preta are important elements of regenerative agriculture. Simply put, biochar is charcoal created through the thermal breakdown of biomass, or pyrolysis. It is a carbon-rich, stable solid that can survive for thousands of years in soil. Several publications substantiate the well-known application of charcoal/ biochar as a soil amendment. The use of charcoal in agriculture was then referred to by a variety of terms, including "agrichar." In 2009, an international scientific conference in Birmingham, England, officially adopted the term "biochar." Biochar improves agricultural productivity, increases soil fertility in acidic (low pH) soils, and offers protection against certain soil and foliar-borne diseases.

In actuality,"Terra Preta"is rich black soil that has been around for hundreds of years. Despite varying weather patterns, these soils have remained exceptionally fertile in many tropical regions. These soils had double crop yields when compared to nearby soils and held roughly three times as much nitrogen and phosphorus, with only 9% carbon. They appeared to be permanently fertile. Regeneration of terra preta from forest litter suggests that the microbiome of terra preta may have as much of an influence as added charcoal, imparting the character of the underlying material on organic material that decays upon it. Terra preta has even been found to promote mycorrhiza and host higher populations of diazotrophs (nitrogen-fixing bacteria)

7. Intercropping: Even though intercropping and multiple cropping have no documented histories, planting crops together has a long history based on the evidence currently available. A multiple cropping system known as intercropping involves planting two or more crop species in a field at the same time during the growing season. Naturally, this does not imply that plants can be planted together at the same time during intercropping; rather, the goal is to have two or more crops together in one location for the duration of the growing season, or at least a portion of it. Many reports concerning the positive effects and superiority

of intercrop than pure crop are increasing and stable productivity, maximum use of environmental resources, reduces pest, diseases and weeds and also improve fertility. The intercropping is divided into the following four groups (Vandermeer, 1992; Ofori and Stern, 1987):

i) Row-intercropping: Growing two or more crops simultaneously where one or more crops are planted in regular rows, and other crops may be grown simultaneously in rows or randomly with the first crop.

ii) Mixed-intercropping: Growing two or more crops simultaneously with no distinct row arrangement. This type can be suitable for grass-legume intercropping in pastures.

iii) Strip-intercropping: Growing two or more crops simultaneously in different strips wide enough to permit independent cultivation but narrow enough for the crops to interact ergonomically.

iv) Relay-intercropping: Growing two or more crops simultaneously during part of the life cycle of each. A second crop is planted after the first crop has reached its reproductive stage but before it is ready for harvest.

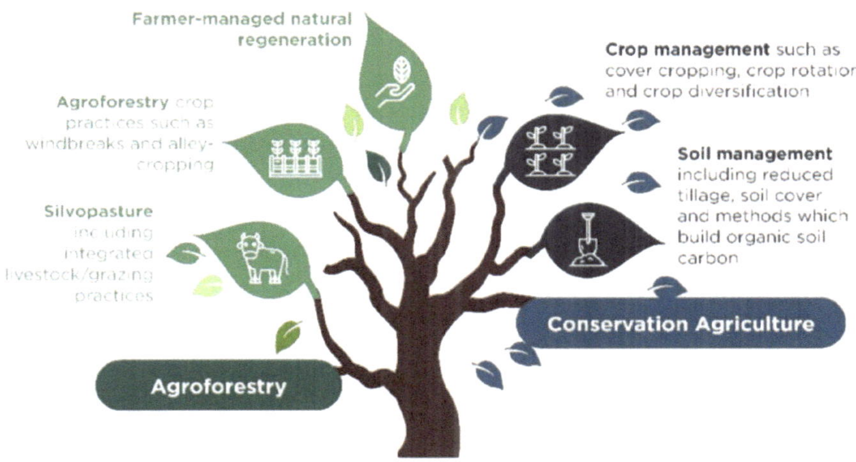

Fig. 2: Regenerative agricultural practices

Benefits of Regenerative Agricultural Practices

By adopting regenerative agriculture practices, the following benefits can be achieved, which are significant over sustainable agriculture:

1. Climate change in the opposite direction: Capturing atmospheric carbon in the soil and ground vegetation is known as carbon sequestration.

When this occurs, it has the potential to help reverse current worldwide patterns of rising greenhouse gas levels in the atmosphere and sea. Improving soil health and planting perennials are two of the most effective approaches to aid soil carbon sequestration.

2. Restore the health of watershed organisms: When it rains, healthy top soil collects the water, while insufficient soil is too dry to do so, causing runoff. This drainage is dangerous because it contains any toxic pesticides or fertilisers that have been dumped into rivers, streams, and the ocean. We can all have access to clean, safe water by encouraging the growth of organic soil matter and avoiding the use of chemicals.
3. Provide Nutrient-Dense Food: Regenerative agriculture practitioners claim that higher yields have resulted in decreased nutrients in plants, and that adopting regenerative agriculture practices can improve the nutritional quality of food. Climate, particularly elevated CO2, can have a negative impact on the nutritional quality of grain
4. Build resistant to climate change: A healthy top soil is resistant to droughts and climate change because it has a high organic matter content that helps retain moisture until it is needed. In addition to helping to keep moisture in the soil, cover crops enable plants to rely on their water stores in times of limited rainfall.
5. Expand crop yield: Many people believe that organic farming produces lower yields, but data seems to indicate otherwise. The Rodale Institute Farming System Trial has shown that organic crops grown without the use of industrial pesticides or genetically modified organisms are more resilient to unfavorable weather. During a drought, for example, yields from organic systems were up to 24–34% higher.
6. The benefits for soil, biodiversity and the Planet: Regenerative agriculture can help: re establish the fertility of the land, by significantly increasing the organic carbon available in the soil and its nitrogen-based components; strengthen soil and plant root structures, thereby limiting erosion and reducing the probability of catastrophic environmental events; increase local biodiversity, not merely by reintroducing now-forgotten crops but also by encouraging local species to grow spontaneously and wildlife to return (Nath *et al.*, 2023b).
7. Improve Prosperity: Regenerative agriculture improves long-term farmer livelihood through reduced costs, improved crop yield and quality, and greater resilience to market volatility and extreme climate events.

Conclusion

Regenerative agriculture is an outcome-based system of food production that improves farm productivity and profitability while preserving biodiversity, the climate, and water resources. Regenerative farming practices enhance the soil's capacity for production as well as the surrounding environment. Non-chemical techniques may help boost the biological potential of the soil environment by lowering or eliminating the use of chemical pesticides and external fertiliser sources. By making sensible use of natural resources, the application of some of the above-mentioned practices could assist nations in becoming more self-sufficient in terms of food supply. A variety of choices for farmers and a more economical use of limited production resources in agriculture will define tomorrow's development strategies.

Regenerative agriculture practices enable farmers to take long-term action by contributing to a larger solution to the crisis through carbon sequestration, in addition to helping them deal with the immediate effects of climate change by making their farms more resilient and adaptive to what is happening around them. To determine whether regenerative agriculture techniques enhance soil biological features and fertiliser efficiency and, consequently, lessen dependency on synthetic inputs, long-term multidisciplinary research is required. both the government and business unlocking this potential and creating cutting-edge, economically viable regenerative farming technology suitable to Mediterranean climates, along with extension initiatives that boost its acceptance and implementation while guaranteeing food and nutritional security, all depend on research support. The integration of these practices not only enhances agricultural productivity but also contributes to the overall resilience and sustainability of our ecosystems.

References

Balfour, E. B. (1943). The Living Soil. London, UK: Faber and Faber.

Coleman, D. C., & Crossley Jr., D. A. (1996). Fundamentals of Soil Ecology. Academic Press, San Diego, CA, 205 pp.

Dayal, P., Kumar, A., Singh, M., Sow, S., Ranjan, S., Pal, R.K., Tyagi, S., and Jena, T. (2023). Soil fertility status and physio-agronomic performance of wheat grown under contrasting tillage and weed management scenarios. Journal of Soil and Water Conservation, 22(3), 242-249.

Giller, K. E., Hijbeek, R., Andersson, J. A., & Sumberg, J. (2021). Regenerative agriculture: an agronomic perspective. Outlook on agriculture, 50(1), 13-25.

Howard, A. (1943). An agricultural testament. Nature, 147(3733), 590-591.

Howard, A., & Howard, L. E. (1945). Farming or Gardening for Health or Disease (Published later as The Soil Health: A Study of Organic Agriculture, 2010.) London, UK: Faber and

Khangura, R., Ferris, D., Wagg, C., & Bowyer, J. (2023). Regenerative Agriculture—A Literature Review on the Practices and Mechanisms Used to Improve Soil Health. Sustainability, 15(3), 2338.

Kumar, N., Sow, S., Rana, L., Singh, A.K., Kumar, A., Kumar, A., and Singh, S.N. (2023). Physio-agronomic performance of sugarcane (Saccharum spp. hybrid complex) genotypes under various planting geometry. Ann. Agric. Res. New Series, 44(1), 93-98.

Lal, R. (2020). Regenerative agriculture for food and climate. Journal of soil and water conservation, 75(5), 123A-124A.

Nath, D., Suman, S.N., Raigar, B. L., Sow, S., and Ranjan, S.R. (2023b). Carbon Dynamics in Soil in Relation to Climate Change. In: Climate Smart Approaches towards Sustainable Crop Production, Vital Biotech Publication, pp. 89-118.

Nath, D., Suman, S.N., Sow, S., and Ranjan, S.R. (2023a). Use of Polyhalites as a Novel Technique to Enhance Nutrient Use Efficiency. In: Innovations in Nutrient Management: Improving Efficiency and Crop Production, Elite Publishing House, pp 50-70.

Pretty, J. N. (1995). Regenerating Agriculture: Policies and Practice for Sustainability and Self-reliance. Earthscan Publications, 320 pp.

Ranjan, S., and Sow, S. (2021). Fertigation: An efficient means for fertilizer application to enhance nutrient use efficiency. Food and Scientific Reports 2(5), 24-27.

Ranjan, S., Kumar, S., Dutta, S.K., Padhan, S.R., Dayal, P., Sow, S., Roy, D.K., Nath, D., Baral, K., and Bharati, V. (2023a). Influence of 36 years of integrated nutrient management on soil carbon sequestration, environmental footprint and agronomic productivity of wheat under rice-wheat cropping system. Front. Environ. Sci., 11, 1222909.

Ranjan, S., Kumar, S., Dutta, S.K., Sow, S., Kumar, S., and Sushant. (2023b). Long-Term Organic Amendment Application Improves Soil Fertility Status, Nutrient Accumulation and Crop Productivity Under Rice- Wheat Cropping System. Communications In Soil Science and Plant Analysis, 54(18), 2579-2589.

Rodale, R. (1983). Breaking new ground: the search for a sustainable agriculture. The Futurist, 1, 15–20.

Roy, D. K., Sow, S., and Ranjan, S. (2023a). Productivity and profitability of zero till winter maize as influenced by integrated weed management practices. Indian Journal of Weed Science, 55(3), 264-268.

Roy, D. K., Sow, S., and Ranjan, S. (2023b). Weed management effect on weeds, productivity and economics of soybean. Indian Journal of Weed Science, 55(2), 228-230.

Sadhu, S., Sow, S., Ranjan, S., and Sahoo, R. (2023). Polymer Coated Fertilisers and Their Scope in Modern Agriculture. In: Current Trends and Advances In Agricultural Sciences, Kripa-Drishti Publications, pp 32-41.

Sahoo, R., Sow, S., Ranjan, S., Samanta, D., and Sadhu, S. (2023). Effect of Climate Change in Plant Disease Development and Its Management Strategies. In: Recent Fronteers in Plant Science Research, Blue Duck Publications, Srinagar, J&K, pp 34-42.

Schreefel, L., de Boer, I. J. M., Timler, C. J., Groot, J. C. J., Zwetsloot, M. J., Creamer, R. E., *et al.* (2022). How to make regenerative practices work on the farm: A modelling framework. Agricultural Systems, 198, 103371.

Shao, Z., Mwakidoshi, E.R., Muindi, E.M., Soratto, R.P., Ranjan, S., Padhan, S.R., *et al.* (2023). Synthetic Fertilizer Application Coupled with Bioslurry Optimises Potato (Solanum tuberosum) Growth and Yield. Agronomy, 13, 2162.

Shaxson, T. F. (1997). Conservation at the crossroads in tropical countries (Commentary). J. Soil Water Conser. Soc., January–February, p. 2.

Sow, S., and Kumar, N. (2023). Application of remote sensing in sugarcane cultivation. In: Sustainable Sugarcane Production and Utilization: Issues and Initiatives, Dr. Rajendra Prasad Central Agricultural University, Pusa, Bihar, pp 202-206.

Sow, S., and Ranjan, S. (2021). A Review on Agronomic Biofortification for Improving Food and Nutritional Security. Research Biotica, 3(2), 139-144.

Sow, S., Ranjan, S., Behera, B., Ghosh, M., Kumar, S., and Kumar, S.K. (2023a). Maintaining agricultural sustainability through carbon footprint management. Current Science, 125(9), 939-944.

Sow, S., Ranjan, S., Ghosh, M., Kumar, S., and Choudhury, S.R. (2020). Crop Residues as Potential Sustainable Source of Silica in the Era of Climate Change: A Review. International Journal of Current Microbiology and Applied Sciences, 10(11),1629-1641.

Sow, S., Singh, G., Ghosh, M., Dutta, S.K., Mandal, N., Kumar, S., and Ranjan, S. (2023b). Nitrogen-management strategy through leaf-colour chart and SPAD meter for optimising the productivity in irrigated wheat (Triticum aestivum). Indian Journal of Agronomy, 68(2), 219-222.

Stockdale, E. A., Lampkin, N. H., Hovi, M., Keatinge, R., Lennartsson, E. K. M., *et al.* (2001). Agronomic and environmental implications of organic farming systems. Advances in Agronomy, 70.

16

Water Conservation Techniques for Sustainable Crop Production

Bibek Ishore and Tanmoy Das

Department of Farm Machinery and Power Engineering, College of Agricultural Engineering and Technology, Dr. Rajendra Prasad Central Agricultural University, Pusa, Bihar 848125, India

Abstract

The availability of water for irrigation is decreased by a lack of resources. Freshwater scarcity demands efficient ways of growing crops which in turn requires effective water management. In order to increase the overall performance and productivity of water use, concepts and performance descriptors that may be helpful in defining water conservation and saving of water are explored. Synchronisation of available water use and its resources can maintain sustainable agricultural production, ensuring a consistent water supply dividend. Proper farm management of a watershed based on its water availability with respect to farm operations by scheduling on and off-farm transportation or supply of water and accurate irrigation water utilisation can improve the sustainability of available water for consumptive use. Precise application of irrigation water is not only an option to conserve water but also reduces the chances of unfavourable water losses during application.

Keywords: *Crop production, On and off farm transportation, Water productivity, Water saving*

Introduction

Water is a prominent element in the daily life of human beings as well as in agriculture. In better crop production, the quantity of water, but the quality of water is also important. With respect to the current scenario of water harvesting or water conservation there are some techniques adopted in Indian agricultural practices, such as the use of drip irrigation to minimise water surplus, rain-fed agricultural practices, storing rain water in different storage components, using mulching (either natural or plastic), proper tillage practices, selecting crops which need less amount of water, using laser leveller for better water uses to encounter surplus water flow to the lower elevated portion of the field etc. Water conservation techniques and measures are directly related to the type of meteorological factors of specific watershed. For the arid and semi-arid regions proper method of tillage practice improves water use while in conventional practices of farming consumes more than required water quantity. Some crops are water-intensive where the quantity of the available water is crucial. Sources of water can be ground water, rain water and irrigation water. In sustainable crop production, the right amount of water is necessary for the optimal physiological growth of specific crops. Although the availability of water depends upon the geographical location, for example in the Gangetic plains of India there is less need for water conservation agricultural purposes. On the other, in the region of Deccan Plateau of India, availability of water is comparatively lower than in the Gangetic plains. It is also important that water conservation techniques can be adopted irrespective of the geographical locations. Water is conserved in nature itself but the availability of water depends on the hydrological cycle. Therefore, conservation has become a trending topic. Water conservation can be improved through sustainable agricultural practices such as proper scheduling of irrigation, selection of crops based on sustainability, monitoring soil moisture content, maintaining records of crop growth and physio-chemical data, implementing effective drainage systems, intercropping operations and utilising recycled water for other purposes. Water conservation structures like roof top water harvesting, farm ponds, bunding and terracing can help retain rain water to reduce the need for irrigation. Excess water can be managed by farmers through recycling and diverting it to nearby cultivable land.

Water Utilisation

- Efficiency is a common term used to describe the operation of water supply systems and water consumption activities. Nonetheless, there are no commonly agreed definitions, and the term "efficiency" is employed with varying interpretations, mainly in relation to different water usage sectors. Water use efficiency is sometimes synonymous with both water conservation and water saving (Gardu˜no *et al.*, 1994). More conceptual consistency is needed to improve comprehension of the language used in connection with water use performance. When referring to irrigation systems, the term "efficiency" is frequently used to describe each of the following parts: preservation, transportation, off- and on-farm transportation, and on-farm application (Bos *et al.*, 1982 and Wolters, 1992). It can be defined as the ratio of the entire water level delivered to the sub-system in question and the amount of water that the sub-system uses to its advantage. This ratio is typically represented in percentage terms. When calculating on farm treatment productivity, the total amount of water applied to that field is the denominator, and the quantity of freshwater added to the root zone storage is the substitute for the numerator.

- There is a need to develop innovative ideas that provide a clear distinction between useful and unfavourable uses, as well as consumptive and non-consumptive uses. Likewise, the distinctions between the portions of non-consumed water channelled into a watering system or subsystem that are recyclable and non-reusable are being made clearer (Willardson *et al.*, 1994; Allen *et al.*, 1997; Pereira *et al.*, 2002; Perry *et al.*, 2009). In keeping with this, Foster and Perry (2010) also suggested an order of water usage portions with the goal of encouraging responsible freshwater use in agriculture. When used in national watershed management for the creation of freshwater-saving and preservation policies and initiatives, these descriptors' alternative performance metrics are far more pertinent than "watering effectiveness. Producers that use irrigation know these signs better than effectiveness, which is an anticipated benefit.
- In an effort to make water usage intentional, it's critical to identify both useful and unfavourable water uses from the standpoint of water economy. Solomon and Burt (1999), who introduced the term watering ability, also made an attempt at this. The advantageous uses of plant and landscaping watering are those that directly result in an agricultural product or a beautiful garden, grass, or golf course, where the desired outcome may be the preservation of specific biological properties. Uses that result from excessive watering, inadequate distribution system management, or water misuse are considered unfavourable.

Favourable Index for Water Consumption

- Water use signs can be stated for any mechanism or procedure that uses water, whether for irrigation or non-irrigation, and it is possible to make sure that actions are taken to use water more effectively, i.e., to improve results from the standpoint of conserving the availability of water. Organising and handling water supplies during times of shortage may benefit from these parameters. They can be paired with process signs, such as those that deal with the water supplies' quality of service. It is important to stress that in order to accurately determine the efficiency and effectiveness of each part of the system, water circulation must be measured at multiple places throughout the entire structure (Pereira *et al.*, 2012).

- There is a current movement in irrigation to emphasise the need to increase the water productivity (Molden *et al.*, 2003, 2010; Oweis and Hachum, 2003; Clemmens and Molden, 2007). The focus of attention has shifted from the effectiveness of irrigation to water productivity. Nonetheless, Molden *et al.* (2003, 2010) and Steduto *et al.* (2007) have addressed how this term is employed differently in reference to different scales (Fig.1). This analysis focuses solely the yield of water from crops that receive watering. A general definition of water output for farming and landscaping irrigation is the ratio, stated in kg/m3, between the actual crop productivity attained (Ya) and the quantity of water consumed. Because irrigating lawns, gardens, or golf courses gives qualitative yields, observers and consumers of a specific landscape must choose a suitable meaning of Ya. The term "denominator" can relate to either the quantity of irrigation water use or the total usage of water which includes precipitation.
- Freshwater yield must be taken into account from both an economic and natural standpoint. Freshwater has high commercial significance, as well as the financial gain that comes from using it for watering for growers. Perry *et al.* (1997) provided an excellent explanation of the necessity of considering the communal or civic value of water . Other areas of the water user industry also recognise that water improves performance. It needs to be customised to the unique requirements of every industry and operation. It is likely necessary to use or define the phrase "access to water output" differently for every manufacturing or service activity (Pereira *et al.*, 2012).

Agricultural Water Conservation Measures

Soil conservation means protecting the soil from erosion losses and dwindling soil fertility due to acidification, salinisation, intensive agricultural practices and some soil pollutants. Soil loss by water and wind is the greatest evil of agriculture and animal husbandry in India. Farmers applying soil and water conservation practices benefit more in terms of higher yields and conservation of natural resources than the farmers not practicing soil and water conservation.

i) Crop rotation
ii) Strip cropping
iii) Contour ploughing
iv) Tumbling shifting cultivation
v) Contour cultivation
vi) Intercropping

vii) Organic farming
viii) Conservation tillage
ix) Mulching
x) Zabo system
xi) Terracing
xii) Contour bunding
xiii) Continuous contour trenches

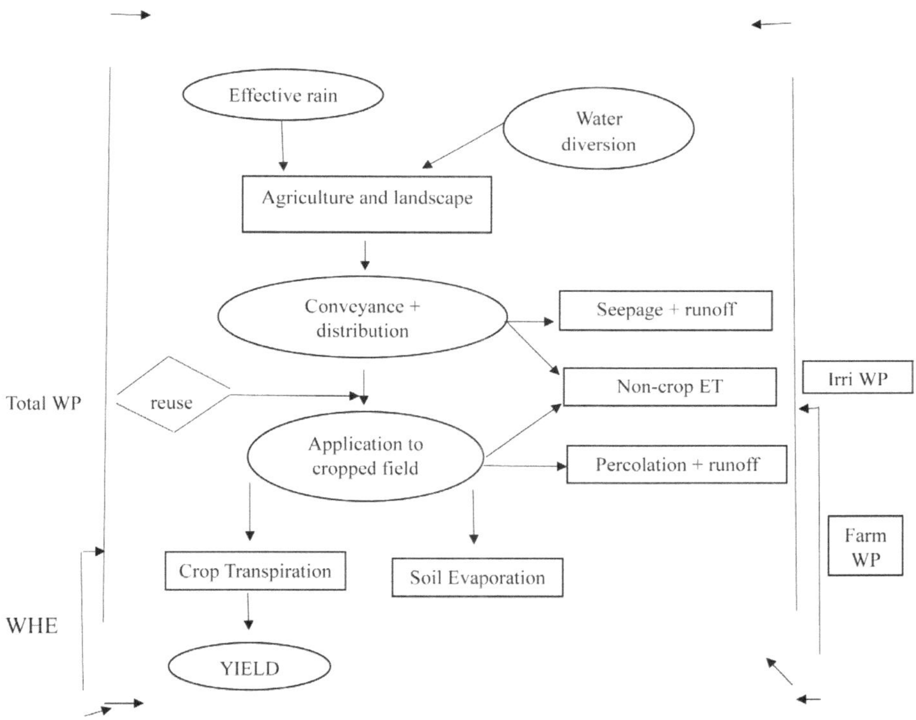

Fig. 1. Illustration of water usage profitability of farming at different categories: (a) the plant's productivity through its water utilisation effectiveness; (b) the watered plant at the farming level; (c) the watered plant at the entire system level and (d) the plant by implementing irrigated water and precipitation. (Source: Pereira *et al.*, 2012)

Chart for Estimating Water Protection

- According to Sarwar *et al.* (2001), First, irrigation schedules with lower water treatments were established for a typical year. Based on 15 years' worth of yearly rainfall records (1980–1994), the mean year was chosen, with a yearly rainfall average of 375 mm. The closest rainfall, 363 mm in 1987 (45% likelihood of surpassing with a 2.54-year return time), was measured in that year. Using the Priestley and Taylor (1972) approach, the standard evapotranspiration ET0 was computed using the daily weather information (i.e., highest and smallest conditions, velocity of the wind, and daylight durations) of 1987. Our main goal is to increase the production of crops while using the least amount of channel reservoirs possible, so it's imperative that we preserve a respectable relative evaporation (Tact/Tpot). Tpot is the possibility of plant evaporation, and Tact is the amount of crop evaporation that has already occurred. Tact / Tpot represents the general conditions in the area not saturated and their impact on plant development since it considers the influence of both the pH of the soil and wetness. Given the groundwater-scarce nature of the FDP area, a certain water restriction is probably present.

- According to Sarwar *et al.* (2001), To forecast the impact of the timing of watering listed in Table 3 on plant water loss, salt content of the soil, water level conduct, and discharge requirements, simulations for a long time (1980–1994) were run. Three distinct irrigation water qualities were taken into consideration because farmers in the FDP area use tubewell water for watering: (1) watering using high-quality channel water, (2) watering using a combination of waterway and tubewell water, and (3) watering utilising tubewell water exclusively. Two scenarios related to water levels in the FDP region were replicated. The first scenario shows areas with a sufficient groundwater channel in place to maintain the level of the water surface at the appropriate elevations. According to Sarwar *et al.* (2000), the outlet thickness is set at 2.2 m and the outlet interval is 500 m. The second scenario reflects the regions with inadequate and salinized freshwater and a failing drainage network. In this case, the level of water was first estimated to be 2 metres below the surface and was permitted to change during the period of growth based on precipitation, crop water consumption, and irrigation quantity.
- According to Sarwar *et al.* (2001), The ongoing models showed that an extremely low water table has a significant impact on tubewell water usage. Both capillary circulation and tubewell watering are used to add minerals. Under extremely humid circumstances, plants have limited root water uptake, which causes infiltrated water to drive the extremely acidic water level upward. This process produces wet conditions in addition to raising the pH level of the underground layer. It will not be possible to leach minerals from these bottom saltwater table places using subpar irrigation water, which will cause the lands to lose output even more quickly.

Economic Importance

- According to Ward *et al.* (2008) the farming assessment is based on calculating the income-optimised planting strategies adjustments in response to various drip-watering subsidies. Estimating how plant and watering technology-related region output responds to different drip watering investment assistance, which range from 0% to 100%, is the foundation of the economic assessment of water. In this region, drip farming is significantly more expensive than flood watering, as is typical around the world. It also offers higher agricultural yields with a smaller amount of water utilised per acre. The response to whether drip farming is more practical for growers depends on what mix of factors related to the economy and water availability makes drip watering more beneficial than flood watering.
- According to Ward *et al.* (2008) the basin-wide framework identifies trends in water usage and water-related actions that optimise net advantages' depreciated current worth. The framework was developed to find trends in water usage that optimise the anticipated current value of financial advantages in terms of water applications, regions and duration periods of time. The aforementioned farm income is reflected in the overall basin-scale net advantages. For utilisation in agriculture, downtown areas, and the natural environment, total advantages are specified.
- Ward *et al.* (2008) stated that more expenses are usually needed to make the extra water used suitable for consumption by people when waterways become more channelled or limited. The biggest additional costs for commercial water supply pumping networks are those connected to electricity and related upkeep and operation. Variable and fixed expenses are the two categories of expenses.
- Ward *et al.* (2008) stated that the expense of water treatment in large cities to make it healthy and secure for people to ingest is high. Residential treatment expenses are often lower for water being pumped than for rerouted water from rivers, although they are still considerably higher than for farming. Information on farm water expenses was obtained from official farm corporate expense and profit plans, whereas data on urban transportation expenses came from the water utilities in El Paso and Albuquerque. When expenses are subtracted from benefits, both city and environmental expenses are incorporated as negative elements in the desired value.

Conclusion

The current methods of watering used by growers are designed to supply the most water possible in order to optimise crop yield. It has been demonstrated that taking into account commercial water usage criteria is important when analysing water supply and operations upgrades at the farm and scheme sizes. The real-world implementation of these commercial water usage indicators can be seen through scenarios. It is reasonable to assume that their acceptance will lead to a greater awareness of water preservation and protection. Comparable ideas of water profitability can be applied to non-irrigation uses, taking into account the expenses associated with effluent discharge and treatment—two key issues in water preservation and conservation. It is anticipated that adopting an innovative transparency in weighing every advantage and disadvantage as well as more exact terminology use could significantly enhance connections across disciplines and help achieve significant gains in the management of limited assets. Decision-makers in the government and water agencies, as well as producers and consumers in general, may inevitably gain a better understanding of the productivity and performance of water.

References

Bos M.G. and Nugteren J. (1982). On Irrigation Efficiencies. Int'l Inst. for Land Reclamation and Improvement, ILRI Publ. 19: 1- 138.

Clemmens A.J. and Molden D.J. (2007). Water uses and productivity of irrigation systems. Irrig. Sci. 25: 247–261.

Foster S.S.D. and Perry C.J. (2010). Improving groundwater resource accounting in irrigated areas: a prerequisite for promoting sustainable use. Hydrogeol. J. 18: 291–294.

Gardu˜no H and Arreguín-Cortés F (Eds.). (1994). Efficient Water Use. UNESCO Regional Office ROSTLAC, Montevideo, Uruguay: 1-379.

Molden D., Murray-Rust H., Sakthivadivel R. and Makin, I. (2003). A water-productivity framework for understanding and action. In: Kijne, J.W., Barker, R., Molden, D. (Eds.), Water Productivity in Agriculture: Limits and Opportunities for Improvement. IWMI and CABI Publ., Wallingford, 1–18.

Molden D., Oweis T., Steduto P., Bindraban P., Hanjra M.A. and Kijne J. (2010). Improving agricultural water productivity: between optimism and caution. Agric. Water Manage. 97: 528–535.

Oweis T.Y., Hachum A.Y. (2003). Improving water productivity in the dry areas of West Asia and North Africa. In: Kijne, J.W., Barker, R., Molden, D. (Eds.), Water Productivity in Agriculture: Limits and Opportunities for Improvement. IWMI and CABI Publ., Wallingford:179–198.

Pereira L.S., Cordery I. and Iacovides I (2012). Improved indicators of water use performance and productivity for sustainable water conservation and saving. Agricultural Water Management 108: 39– 51.

Pereira L.S., Oweis T. and Zairi A. (2002). Irrigation management under water scarcity. Agric. Water Manage. 57: 175–206.

Perry C, Steduto P, Allen R.G. and Burt C.M. (2009). Increasing productivity in irrigated agriculture: agronomic constraints and hydrological realities. Agric. Water Manage. 96 (11):s 1517–1524.

Perry C.J., Rock M. and Seckler D. (1997). Water as an economic good: a solution, or a problem? Research Report 14. International Irrigation Management Institute, Colombo, Sri Lanka, 16 pp.

Sarwar A. and Bastiaanssen W.G.M. (2001). Long-term effects of irrigation water conservation on crop production and environment in semiarid areas. J. Irrig. Drain Eng.,127:331-338.

Sarwar A., and Feddes R. A. (2000). ''Evaluating drainage design parameters for the Fourth Drainage Project, Pakistan by using SWAP model. Part II—Modeling Results.'' Irrig. and Drain. Sys., Dordrecht, The Netherlands, 14: 281–299.

Solomon K.H. and Burt C.M. (1999). Irrigation sagacity: a measure of prudent water use. Irrig. Sci. 18: 135–140.

Steduto P., Hsiao T.C. and Fereres E. (2007). On the conservative behaviour of biomass water productivity. Irrig. Sci. 25: 189–207.

Ward F.A. and Velazquez M.P. (2008). Water conservation in irrigation can increase water use. The National Academy of Sciences of the USA, 47: 18215–18220.

Willardson L.S., Allen R.G. and Frederiksen H.D. (1994). Universal fractions and the elimination of irrigation efficiencies. In: Proceedings 13th Technical Conference, US Commission on Irrigation and Drainage:1-18.

Wolters W. (1992). Influences on the Efficiency of Irrigation Water Use. ILRI Publ. 51: 1-79.

17

Weather Variability and Crop Yield Forecasting for Sustainable Agriculture

Hemareddy Thimmareddy[1], Nikhil S Paschapur[2], A.C. Mahadevaprasad[3], Vishwanath V. Meti[3], R. Divya[3] and Yashwanth H.R.[3]

[1]*Agro Climate Research Centre, Tamil Nadu Agricultural University Coimbatore Tamil Nadu, India*

[2]*Department of Climate Change and Agricultural Meteorology, Punjab Agricultural University, Ludhiana, India*

[3]*Department of Agricultural Meteorology, University of Agricultural Sciences Dharwad, Karnataka, India*

Abstract

The Intergovernmental Panel on Climate Change (IPCC) presents compelling evidence of rising temperatures and disrupted precipitation patterns. The urgency is underscored as agriculture faces vulnerabilities, with uncertainties in planting times, water management, and pest control due to increased weather variability. Climate change, driven by anthropogenic activities like fossil fuel combustion and deforestation, intensifies with unprecedented greenhouse gas emissions. Extreme events, including heatwaves, floods, droughts, and storms, accentuate the vulnerabilities of agriculture. Mitigation efforts to reduce emissions are crucial, but adaptation strategies, especially in agriculture, are equally vital. Accurate climate models and forecasting play a pivotal role in building resilience and ensuring food security, demanding a deeper understanding of the intricate interactions between human activities and atmospheric processes. This chapter highlights advanced meteorological techniques fuelled by technological evolution, including remote sensing, high-performance computing, Doppler radar, Lidar, climate modelling, and artificial intelligence. These technologies contribute to precise weather forecasting, climate studies, and disaster management. In the realm of precision agriculture, the significance of crop yield forecasting models is examined. These models leverage historical and current data, climate information, and advanced analytics to empower farmers, policymakers, and stakeholders. Meteorological data integration, machine learning, and remote sensing methods enhance the accuracy of these models. Global case studies underscore the relevance of forecasting models in adapting to variable climatic conditions. The Clean Development Mechanism under the Kyoto Protocol is introduced as a crucial initiative, offering insights into its objectives and operational details in India. Implications for sustainable agriculture include climate resilience, judicious water management, and evidence-based policy formulation.

Keywords: *Climate Change, Extreme Events. IPCC, Yield Forecasting*

Introduction

Climate change has brought about unprecedented weather variability, disrupting established patterns and posing significant challenges to global agriculture. The Intergovernmental Panel on Climate Change (IPCC) reports unequivocal evidence of increasing temperatures and altered precipitation patterns (IPCC, 2021). These changes impact crop production, water availability, and overall food security, necessitating innovative approaches to ensure sustainable agriculture in the face of evolving climatic conditions. To address these challenges, accurate crop yield forecasting becomes paramount. Reliable predictions empower farmers, policymakers, and agricultural stakeholders to make informed decisions regarding resource allocation, risk management, and adaptation strategies (Lobell *et al.*, 2012; FAO, 2019). This chapter explores the intricate interplay between weather variability, climate change, and the development of effective crop yield forecasting models, offering a comprehensive perspective on how sustainable agriculture can be achieved in the midst of these dynamic environmental shifts.

Climate Change and Weather Variability

Climate change, driven by anthropogenic activities, has emerged as one of the most pressing challenges of our era, profoundly influencing weather patterns worldwide. The Intergovernmental Panel on Climate Change (IPCC) affirms that human activities, notably the combustion of fossil fuels and deforestation, have led to unprecedented levels of greenhouse gas emissions, intensifying the warming of the Earth's atmosphere (IPCC, 2021). This global warming sets in motion a cascade of effects, prominently manifested in the form of weather variability.

Weather variability, the deviation from the average weather conditions over a specific period, has become more pronounced and erratic due to climate change (Seneviratne *et al.*, 2012). Extreme events such as heatwaves, floods, droughts, and storms are becoming more frequent and intense, disrupting established climate norms (Ranjan *et al.*, 2021). These variations pose significant challenges to various sectors, with agriculture being particularly vulnerable. Agricultural productivity is intricately linked to climate and weather patterns. Changes in temperature, precipitation, and the frequency of extreme events directly impact crop growth, water availability, and overall agricultural yield (Wheeler and von Braun, 2013). Farmers face uncertainties in planting times, water resource management, and pest control, making traditional farming practices increasingly untenable.

Addressing climate change and weather variability necessitates a multi-faceted approach. Mitigation efforts to reduce greenhouse gas emissions are imperative to curb the pace of climate change. Concurrently, adaptation strategies, informed by accurate climate models and forecasting, are crucial for sectors like agriculture to build resilience and ensure food security (Rosenzweig *et al.*, 2014; FAO, 2018). As we navigate these uncharted territories of climate change and weather variability, a deeper understanding of the complex interactions between human activities, atmospheric processes, and their cascading effects on weather patterns is essential. This note explores the critical dimensions of this dynamic relationship, emphasising the urgency for global collaboration and innovative solutions to mitigate and adapt to the changing climate.

Advanced Meteorological Techniques

Meteorology, the study of the Earth's atmosphere and its phenomena, has undergone a transformative evolution propelled by technological advancements. The integration of cutting-edge technologies into meteorological practices has ushered in a new era of precision and depth in our understanding of atmospheric dynamics. This note delves into the realm of advanced meteorological techniques, exploring their pivotal role in enhancing weather forecasting, climate studies, and disaster management.

a. **Remote Sensing and Satellite Technology:** Advanced meteorological techniques leverage remote sensing and satellite technologies to capture a comprehensive view of the Earth's atmosphere. Satellites equipped with sophisticated sensors provide real-time data on cloud cover, atmospheric moisture, temperature patterns, and even trace gas concentrations (Vonder Haar *et al.*, 2010). This wealth of information enhances the accuracy of weather predictions and facilitates the monitoring of long-term climate trends.

b. **High-Performance Computing and Numerical Weather Prediction:** High-performance computing (HPC) capabilities have revolutionised numerical weather prediction models. Complex mathematical models, such as those based on the Finite-Volume Cubed-Sphere (FV3) dynamical core, enable meteorologists to simulate atmospheric processes with unprecedented detail (Putman *et al.*, 2014). The synergy of HPC and advanced algorithms has significantly extended the lead time and accuracy of weather forecasts.

c. **Doppler Radar and Lidar Technology:** Doppler radar and Lidar (Light Detection and Ranging) technologies have enhanced our ability to monitor and track severe weather events. Doppler radar provides detailed information on precipitation intensity, wind patterns, and storm rotation, aiding in the timely issuance of weather warnings (Doviak and Zrnić, 1993). Lidar, employing laser light, contributes to atmospheric profiling and the study of aerosol distribution.

d. **Climate Modeling and Simulation:** Advanced meteorological techniques play a pivotal role in climate modeling and simulation. Complex climate models, such as those employed by the Community Earth System Model (CESM), integrate various components of the Earth system, including the atmosphere, oceans, land surface, and sea ice (Hurrell *et al.*, 2013). These models aid in understanding long-term climate trends and predicting the impacts of anthropogenic activities.

e. **Artificial Intelligence and Machine Learning:** The integration of artificial intelligence (AI) and machine learning (ML) techniques has further augmented meteorological capabilities. AI algorithms analyse vast datasets to identify patterns, improve forecasting accuracy, and optimise decision-making processes (Liu et al., 2021).

Crop Yield Forecasting Models

Agricultural productivity and food security are intricately linked to the ability to predict crop yields accurately. Crop yield forecasting models have emerged as crucial tools in the realm of precision agriculture, providing valuable insights for farmers, policymakers, and stakeholders. This note explores the significance of these models, their methodologies, and the impact they have on sustainable agriculture. Crop yield forecasting models are computational frameworks that leverage historical and current data, climate information, and advanced analytics to estimate future crop yields. These models play a pivotal role in mitigating risks associated with uncertain weather patterns, optimising resource allocation, and enhancing overall agricultural efficiency (Lobell *et al.*, 2019). The yield forecast can be achieved by different methods like:

a. **Meteorological and Climate Data Integration:** Integrating meteorological data into crop yield models is fundamental for accurate predictions. Factors such as temperature, precipitation, and solar radiation significantly influence crop growth and development. Advanced models often utilise remote sensing and satellite technologies to capture real-time climate information (Rötter *et al.*, 2018).

b. **Machine Learning Approaches:** Modern crop yield forecasting models increasingly employ machine learning (ML) techniques. These approaches, including support vector machines, random forests, and neural networks, excel at identifying complex patterns within large datasets, offering enhanced prediction accuracy (Ruan *et al.*, 2020).
c. **Remote Sensing and Satellite Imagery:** Remote sensing technologies, such as satellite imagery, contribute to high-resolution spatial and temporal data for crop monitoring. This information aids in assessing crop health, identifying stress factors, and predicting yield variations. Models that incorporate remote sensing data provide a dynamic and real-time dimension to forecasting.
d. **Data-Driven Decision Support:** Crop yield forecasting models serve as decision support tools for farmers and policymakers. By offering early insights into potential yield outcomes, these models empower stakeholders to make informed decisions regarding planting, irrigation, and resource allocation. This proactive approach contributes to increased resilience in the face of climate variability (Dong *et al.*, 2018).
e. **Impact on Sustainable Agriculture:** The integration of crop yield forecasting models into agricultural practices aligns with the goals of sustainable agriculture. By optimising resource use, reducing waste, and enhancing resilience to climate change, these models contribute to the long-term viability of farming systems (Asseng *et al.*, 2019).

Case Studies

United States - Midwest Drought (2012)

The severe drought in the U.S. Midwest in 2012 showcased the critical role of forecasting models. Institutions such as the National Oceanic and Atmospheric Administration (NOAA) and the U.S. Department of Agriculture (USDA) utilised advanced meteorological techniques to predict the drought's impact on corn and soybean yields, offering early warnings for farmers to adjust strategies.

India - Monsoon Variability (2015)

India's reliance on monsoons was evident in 2015 when erratic patterns affected different regions. The Indian Meteorological Department (IMD) and the Indian Council of Agricultural Research (ICAR) employed crop yield forecasting models, aiding farmers in adapting to variable monsoons and promoting sustainable agriculture (Paswan *et al.*, 2023).

Australia - El Niño and Wheat Production (2007)

Australia faced a below-average wheat production year in 2007 due to El Niño. The Australian Bureau of Meteorology's forecasting models assisted farmers in understanding El Niño's climatic implications, enabling strategic decisions on planting times, crop varieties, and irrigation.

Brazil - Soybean Production (2020)

In 2020, Brazil encountered weather challenges impacting soybean production. Forecasting models from the Brazilian Agricultural Research Corporation (EMBRAPA) provided valuable insights, guiding farmers in adjusting planting schedules and irrigation practices to sustain soybean production.

China - Typhoon Impact on Rice (2018)

China's experience with Typhoon Mangkhut in 2018 highlighted the importance of meteorological models. The China Meteorological Administration's forecasting tools played a crucial role in predicting the typhoon's path, assisting farmers in East Asia to take preventive measures and minimise damage to rice crops.

Implications for Sustainable Agriculture

Climate Resilience: Crop yield forecasting enhances the resilience of agriculture in the face of climate change. By anticipating weather-related challenges, farmers can adopt adaptive strategies, such as crop diversification or altered planting schedules (Asseng *et al.*, 2015).

Water Management: Forecasting models assist in judicious water management by predicting periods of water stress or excess. This information empowers farmers to implement water-saving practices, contributing to sustainable water use (Bastiaanssen *et al.*, 2005).

Policy Formulation: Governments and policymakers leverage crop yield forecasts to formulate effective agricultural policies. These policies can address food security, disaster response, and sustainable farming practices (Brown and Funk, 2008).

Challenges and Future Directions

Integration of Advanced Technologies: Embracing emerging technologies such as artificial intelligence, machine learning, and advanced remote sensing can enhance the predictive power of forecasting models. Integrating these tools allows for more nuanced analysis of complex interactions.

Global Collaboration and Data Sharing: Addressing data limitations requires global collaboration and a commitment to data sharing. Initiatives that promote open data platforms and international cooperation can help overcome disparities in data availability (Fuglie *et al.*, 2019).

Climate-Resilient Crop Varieties: Future forecasting models should account for ongoing efforts in developing climate-resilient crop varieties. Incorporating data on the performance of these varieties under varying conditions enhances the accuracy of predictions (Challinor *et al.*, 2018).

Localized and Contextual Models: Tailoring forecasting models to local contexts and specific crops can improve accuracy. Recognising the diverse challenges faced by different regions and crops ensures that forecasting strategies are relevant and effective (Rosenzweig *et al.*, 2013).

Clean Development Mechanism

In the fight against climate change, the Kyoto Protocol has implemented a clean development mechanism. This agreement between developed and developing countries seeks to reduce emissions to preserve the ozone layer and strive for a cleaner environment (Sadhu *et al.*, 2022). Within this initiative, economically advanced nations provide incentives to developing countries to implement projects that mitigate greenhouse gas emissions, such as carbon dioxide. In return for their investment, these nations earn CER credits or Emission Reduction Units, each equivalent to one tonne of CO2.

What is Carbon Credit?

- The concept of carbon credit exchange seeks to incentivise countries to reduce greenhouse gas emissions by rewarding those who meet their commitments and offering financial incentives to developing nations.
- Surplus credits, obtained by surpassing emission reduction targets, can be traded on the global market.
- Each credit is equivalent to one tonne of CO2 emissions avoided.
- This initiative aims to boost funding for developing nations where numerous emission reduction projects are underway.

Objectives of Clean Development Mechanism

A clean development mechanism project is required to demonstrate measurable advancements in climate change mitigation, offering tangible and enduring benefits with a substantial reduction in greenhouse gas emissions. The key objectives of a clean development mechanism encompass:

- Contributing to the prevention and mitigation of climate change.

- Facilitating the development of sustainable strategies in developing countries.
- Aiding industrialised nations in emission reduction and the transition to eco-friendly energy sources.
- Supporting the implementation of innovative emission reduction strategies.
- Reducing dependence on fossil fuels.
- Harnessing animal excrement for energy generation and actively managing it.
- Minimizing pollution generated during manufacturing processes.

Operating Details of Clean Development Mechanism in India

1. Project Identification: This initial phase involves comprehensive research to identify a concept capable of reducing greenhouse gas emissions.
2. Government Approval: Once the concept is identified, it is submitted to the Ministry of Environment, Forest, and Climate Change for approval by the Indian government.
3. Project Development: Research is conducted to establish a baseline against which changes in emissions will be monitored in alignment with the Kyoto Protocol.
4. Authentication: An impartial body, appointed by the CDM Administrative Entity, verifies the results of the initial identification survey.
5. Registration Process: Formal approval by the governing council transforms the selected project into a CDM project, providing it with all the financial and legal benefits outlined in the Kyoto Protocol.
6. Tracking: Post-registration, changes in greenhouse gas emissions are monitored over time, and necessary improvements to the project's implementation are implemented.
7. Verification: A team of specialists scrutinises all data and results before submission for certification.
8. Certification: Following thorough verification, the supervising authority affirms that the project has effectively reduced emissions in accordance with the plan.

Conclusion

In the face of escalating climate change, there's an urgent call for sustainable agriculture. The Intergovernmental Panel on Climate Change (IPCC) provides compelling evidence, emphasising the need for a paradigm shift. This exploration covers climate change's profound impact, the role of advanced meteorological techniques, and the significance of crop yield forecasting models. Fuelled by anthropogenic activities, climate change disrupts weather patterns, heightening the importance of precise crop yield forecasting. Advanced meteorological techniques, employing remote sensing and artificial intelligence, offer unparalleled insights. Crop yield models, integrating data and machine learning, emerge as vital tools globally. Despite challenges, global collaboration, technology integration, and resilient crop development offer hope for a sustainable agricultural future.

References

Asseng, S., Ewert, F., Martre, P., Rötter, R. P., Lobell, D. B., Cammarano, D., *et al.* (2015). Rising temperatures reduce global wheat production. Nature Climate Change, 5(2), 143-147.

Asseng, S., Martre, P., Maiorano, A., Rötter, R. P., O'Leary, G. J., Fitzgerald, G. J., *et al.* (2019). Climate change impact and adaptation for wheat protein. Global Change Biology, 25(1), 155-173.

Bastiaanssen, W. G. M., Noordman, E. J. M., Pelgrum, H., Davids, G., Thoreson, B. P., & Allen, R. G. (2005). SEBAL model with remotely sensed data to improve water-resources management under actual field conditions. Journal of Irrigation and Drainage Engineering, 131(1), 85-93.

Brown, M. E., & Funk, C. C. (2008). Food security under climate change. Science, 319(5863), 580–581.

Challinor, A. J., Watson, J., Lobell, D.B., Howden, S.M., Smith, D.R., & Chhetri, N.B. (2018). A meta-analysis of crop yield under climate change and adaptation. Nature Climate Change, 8(4), 287–291.

Dong, F., Zhang, G., Lu, J., & Li, K. (2018). Fuzzy competence model drift detection for data-driven decision support systems. Knowledge-Based Systems, 143, 284-294.

Doviak, R. J., and Zrnic, D.S. (1993) Doppler radar and weather observations. Academic Press, San Diego, CA, p. 562.

FAO (2018). The future of food and agriculture. Alternative pathways to 2050, FAO, FAO, Rome. (http://www.fao.org/3/I8429EN/i8429en.pdf).

FAO. (2019). Food and Agriculture Organisation of the United Nations, The State of Food and Agriculture 2019. Moving forward on food loss and waste reduction.

Fuglie, K. O., *et al.* (2019). Global trends in climate-resilient crop yields. Nature Communications, 10(1), 1–10.

Hurrell, J. W., Holland, M. M., Gent, P. R., Ghan, S., Kay, J. E., Kushner, P. J., *et al.* (2013). The community earth system model: a framework for collaborative research. Bulletin of the American Meteorological Society, 94(9), 1339-1360.

IPCC. (2021). Summary for Policymakers. In: Climate Change 2021: The Physical Science Basis. Contribution of Working Group I to the Sixth Assessment Report of the Intergovernmental Panel on Climate Change, (Masson-Delmotte, V., P. Zhai, A. Pirani, S.L. Connors, C. Péan, S. Berger, N. Caud, Y. *et al.* (eds.), Cambridge University Press, Cambridge, United Kingdom and New York, NY, USA, pp. 3–32.

Lakhankar, T., Jones, A. S., Combs, C. L., Sengupta, M., Vonder Haar, T. H., & Khanbilvardi, R. (2010). Analysis of large scale spatial variability of soil moisture using a geostatistical method. Sensors, 10(1), 913-932.

Lobell, D. B., & Gourdji, S. M. (2012). The influence of climate change on global crop productivity. Plant Physiology, 160(4), 1686-1697.

Paswan, A., Kumar, N., & Sow, S. (2023). Performance, nutrient uptake and economics of rainfed rice varieties under different crop establishment techniques. Ann. Agric. Res., 44(2), 141-147.

Putman, A. I., & Carbone, I. (2014). Challenges in analysis and interpretation of microsatellite data for population genetic studies. Ecology and Evolution, 4(22), 4399-4428.

Ranjan, S., Sow, S., & Kumar, M. (2021). Agromet Advisory Services for Empowering Farmers to Mitigate Climate Change. Agriculture Letters, 2(3-4), 42-46.

Rosenzweig, C., Elliott, J., Deryng, D., Ruane, A. C., Müller, C., Arneth, A., *et al.* (2014). Assessing agricultural risks of climate change in the 21st century in a global gridded crop model intercomparison. Proceedings of the National Academy of Sciences, 111(9), 3268-3273.

Rötter, R. P., Appiah, M., Fichtler, E., Kersebaum, K. C., Trnka, M., & Hoffmann, M. P. (2018). Linking modelling and experimentation to better capture crop impacts of agroclimatic extremes—A review. Field Crops Research, 221, 142-156.

Ruan, G., Li, X., Yuan, F., Cammarano, D., Ata-UI-Karim, S. T., Liu, X., *et al.* (2022). Improving wheat yield prediction integrating proximal sensing and weather data with machine learning. Computers and Electronics in Agriculture, 195, 106852.

Sadhu, S., Satdev., Sahoo, R., Sow, S., Samanta, D., & Ranjan, S. (2022). Impacts of Global Climate Change on Soil Organisms. In: Emerging Trends in Biosciences, InfoCapsule, Uttar Pradesh.

Seneviratne, S.I., Nicholls, N., Easterling, D., Goodess, C.M., Kanae, S., Kossin, J., *et al.* (2012). Changes in climate extremes and their impacts on the natural physical environment. In: Managing the Risks of Extreme Events and Disasters to Advance Climate Change Adaptation, Field, C.B., Barros, V., Stocker, T.F., Qin, D., Dokken, D.J., Ebi, K.L., *et al.* (eds.), A Special Report of Working Groups I and II of the Intergovernmental Panel on Climate Change (IPCC). Cambridge University Press, Cambridge, UK, and New York, NY, USA, pp. 109-230.

Wheeler, T., & Von Braun, J. (2013). Climate change impacts on global food security. Science, 341(6145), 508-513.

Xu, Y., Liu, X., Cao, X., Huang, C., Liu, E., Qian, S., *et al.* (2021). Artificial intelligence: A powerful paradigm for scientific research. The Innovation, 2(4), 100179.

Index

A
Actinomycetes 130, 131
Aeration 150, 152, 153
Agrosilvipastoral Systems 188
Agrochemicals 74, 126, 163, 165, 169, 170
Agronomical Fortification 14
Antagonistic Plants 116
Apoplast 115, 118
Artificial Intelligence 1, 2, 3, 6, 10, 154, 156, 208, 213
Artificial Neural Networks 4
Automation 3, 8, 137, 145

B
Bacteria 54, 88, 114, 118, 119, 130, 131, 153, 188
Biochar 177, 178, 188
Biofortification 13, 14, 15, 16, 18, 19, 20
Biostimulation 134, 153
Bioventing 154
Beejamrutha 47
Biofertilisers 159, 164, 165, 166, 170
Biological Control 51, 56, 91
Biodiversity Preservation 41, 67, 89, 108

C
Climate Change 2, 23, 25, 26, 27, 29, 33, 35, 37, 44, 49, 69, 60, 61, 63, 65, 66, 67, 76, 80, 101, 103, 107, 108, 137, 142, 144, 150, 155, 164, 178179, 180, 181, 183, 190, 191, 205, 206, 207, 209, 211, 213
Climate Smart Agriculture 28
Crop Modeling 23, 24-34
Crop Monitoring 5, 10, 209
Carbon Sequestration 29, 64, 65, 149, 178, 185, 196, 190, 191
Crop Simulation 25, 26, 34
Crop Rotation 47, 52, 53, 5, 77, 83, 89, 144, 149, 177, 182, 184, 185, 186, 198

D
Data Driven Farming 138, 142
Decision Support Systems 54, 141, 144
Dehalogenation 131
Digitalisation 2, 137
Diversification of Food Production 102
DNAs 128
Drainage 140, 152, 190, 196, 200
Drones 5, 56, 140, 143, 146, 154, 170
Drought-resistant 102

E
Ecological Footprint 125, 171
Ecosystem Degradation 67, 89
Environmental Policies 59, 60, 61, 63, 64, 68, 69
Environmental Sustainability 30, 48, 49, 59, 66, 67, 69, 70, 88, 96, 103, 109, 139, 142, 144, 150, 170
Environmental Stewardship 83, 90, 96, 150
Export Potential 170

F
Fallowing 78
Farmers' Profitability 109
Farmyard Manure 153
Food Security 2, 16, 19, 20, 31, 36, 38, 45, 46, 48, 49, 56, 60, 65, 66, 67, 85, 103, 107, 109, 110, 111, 147, 150, 179, 205, 206, 207, 210
Fungi 114, 116, 118, 119, 125, 130, 131, 153

G
Genetic Engineering 18
Government Initiatives 35, 36, 104, 109
Greenhouse Gases 33
Geographic Information Systems (GIS) 140
GPS 8, 137, 138140, 144, 155, 169
Green Revolution 15, 86, 101

H
Herbicides 3, 74, 89, 125, 126, 127, 128, 129-135
Herbicide Persistence 125, 131, 135

Hidden Hunger 13, 14, 15, 16, 19, 20
Hydrolysis 95, 116, 130, 131

I
ICT Technologies 1
Indian Agriculture 48, 4984, 94, 107, 108
Indigenous Agricultural Knowledge 73, 74
Intergovernmental Panel on Climate Change 205, 206, 212
Inter-cropping 47, 74
Integrated Pests and Disease Management 52

J
Jeevamrutha 47

L
Leaching 129, 132, 133, 152, 167
Leaf Area Index 27, 139
Land Use Efficiency 29
Legume Cultivation 83, 84, 88, 89, 90, 91, 92, 93, 94, 96
Lidar Technology 208

M
Machine Learning 1, 3, 4, 5, 33, 144, 154, 156, 169, 205, 208, 209, 210, 213
Malnutrition 13, 14, 15, 15, 19, 20, 95, 98, 102, 103, 106, 110, 159, 170
Microbial Decomposition 130, 132
Millennium Development Goals 15
Millets 14, 101, 102, 103, 104, 105, 106, 107, 108, 109, 110, 111
Mixed Cropping 47, 74, 77
Mulching 46, 47, 77, 182, 195, 199
Mycoremediation 153

N
National Food Security Mission 36, 104
National Mission for Sustainable Agriculture 36, 37, 104
Nutrient Management 28, 37, 94, 159, 161, 162, 163, 164, 170, 179
Nutraceutical Properties 95
Nutrient Solubilizing Microbes 165
Nutritional Security 16, 84, 86, 95, 96, 107, 110, 191

O
Organic Agriculture 55, 77
Organic Matter 55, 108, 128, 129, 130, 133, 134, 149, 151, 179, 181, 182, 185, 187, 190

P
Pest Management 26, 51, 52, 53, 55, 56, 63, 90, 93, 125, 155
Photodecomposition 128
Phytoremediation 134, 153
Plant Parasitic Nematodes 113
Policy Frameworks 59, 64
Postharvest Losses 83, 92
Pradhan Mantri Kisan Samman Nidhi 36
Pradhan Mantri Krishi Sinchayee Yojana 38
Precision Agriculture 5, 28, 56, 125, 137, 138, 139, 140, 141, 142, 143, 146, 147, 149, 155, 156, 169, 170, 205, 208
Problematic Soil 149, 150, 153
Pulse Revolution 83, 86

R
Regenerative Agriculture 55, 180, 181, 182, 183, 84, 186, 188, 189, 190, 191
Residue Management 36, 54, 133, 135
Rashtriya Krishi Vikas Yojana 43
Reactive Oxygen Species 188
Regenerative Agriculture 55, 177, 180, 181, 182, 183, 184, 186, 188, 189, 190, 191
Remote Sensing 1, 4, 33, 56, 138, 139, 140, 143, 155, 169, 205, 206, 207, 209, 210, 213
Rice 7, 13, 18, 86, 104, 108, 109, 116, 165, 210

S
Salt Affected Soils 150
Selective Breeding 13
Social Empowerment 110
Soil Fertility 37, 41, 47, 48, 62, 64, 65, 74, 77, 78, 83, 84, 85, 88, 108, 140, 162, 165, 169, 170, 177, 178, 179, 182, 184, 186, 188, 198
Soil Organic Carbon 129, 179, 186
Soil Health 4, 6, 35, 36, 37, 38, 40, 41, 42, 43, 45, 47, 48, 55, 56, 59, 63, 64, 65, 67, 89, 90, 93, 94, 107, 108, 143, 149, 150, 162, 177, 178, 179, 180, 181, 184, 186, 190
Soil Health Card Scheme 37, 41, 43
Surface Runoff 132
Sustainable Agriculture 5, 13, 21, 36, 37, 38, 39, 41, 43, 48, 52, 65, 69, 73, 94, 98, 103,

104, 108, 109, 111, 125, 135, 137, 139, 140, 141, 142, 154, 155, 182, 184, 189, 205, 206, 208, 209, 210, 212
Sustainable Development Goals 15, 68, 79, 159, 178

T
Terracing 74, 196, 199

U
UV light 128

V
Volatalisation 127, 132,
Variable Rate Technology (VRT) 140, 144

W
Water Management 37, 39, 40, 45, 62, 63, 90, 162, 205, 210
Water Productivity 195, 198
Water Use Efficiency 27, 36, 37, 38, 39, 90, 93, 108, 196
Weather Variability 205, 206, 207
Weed Control 83, 88, 89, 90, 125, 126, 127, 129, 130, 133, 182
Wheat 7, 13, 18, 86, 104, 108, 109, 135, 164, 165, 168, 170, 210
Women's Empowerment 65, 110

Y
Yield Forecasting 30, 170, 205, 206, 208, 209, 210, 213

Z
Zero Budget Natural Farming 46, 47, 48,
Zabo System 199